ADVANCED EXAMPLES IN PHYSICS

By A. O. Allen

SECOND EDITION

PREFACE

CONSIDERABLE difficulty has been met with in the Degree courses in Physics at Leeds University in setting numerical examples of a fairly advanced type, and I doubt not that the same difficulty will have been felt elsewhere. No teacher can afford to dispense with so important a test of the progress of his students as the working of calculations based on the subject-matter of the lectures; yet there appears to be no collection of suitable exercises published at a moderate price. The dictation of questions to a class absorbs valuable time, and the handing round of loose leaflets is a troublesome process. Moreover the large number of students working privately for the examinations of London University must find themselves at a great disadvantage if, as often happens, they have not easy access to a set of University Calendars. This book is designed to meet the difficulty referred to; most of the questions are taken from old examination-papers, the source being acknowledged by a letter in brackets at the end of each exercise. The London and Victoria Universities are indicated by the letters L. and V. respectively; H. denotes that the paper was one for Honours candidates. When no letter is added, the question is either original or has been taken from the lecture note-books in use at Leeds; for permission to use the latter I am greatly indebted to Professor Stroud. Questions taken from old papers have as a rule been reproduced in their original form even when improvement seemed possible; the detection of such cases is in itself a useful exercise.

A. O. ALLEN.

LEEDS UNIVERSITY.

GENERAL PHYSICS.

1. Find the number of watts in one horse-power, given 1 foot = 30·48 cms. ; 1 lb. = 453·6 gms. ; $g = 981$ cms. per sec. per sec. L.

2. A body oscillates under a restoring force proportional to the displacement. Shew that slight frictional resistance steadily diminishes the amplitude, but does not sensibly affect the period. Shew by dimensions or otherwise that if the restoring force varies as the nth power of the displacement, the period varies inversely as the $\frac{1}{2}(n-1)$th power of the amplitude. L.

3. Justify on theoretical grounds the empirical law that the resistance of a medium devoid of rigidity to the passage of a sphere varies as the density of the medium and the square of the velocity and diameter jointly. Define the "terminal velocity." L.

4. Shew that if a particle be subject to an impressed force g per unit mass, and to a resistance per unit mass equal to k times the velocity, then however quickly or slowly the particle be started, it will eventually move at a certain constant speed in the direction of the impressed force. Determine the magnitude of this "terminal velocity."

5. On the experimental law that the resistance of similar steamers varies as the area of the wetted surface and the square of the velocity, prove that if a 6 ft. model of 0·01 ton displacement run at a speed of 2 knots in an experimental tank experiences a resistance of 0·2 lb., a similar 600 ft. 10,000 ton steamer at 20 knots would experience a resistance equivalent to an incline of 1 in 112, and require over 12,000 effective horse-power.   L.

6. Assuming the experimental laws that the normal pressure of the wind and the tangential friction of the water per sq. ft. are proportional to the square of the velocity, prove that a ship and its model are heeled over to the same angle by a wind proportional to the square root of the length or the sixth root of the tonnage and move through the water at a proportional speed.   L.

7. On the above experimental laws, prove that steamers geometrically similar, propelled at speed proportional to the sixth root of the tonnage, will experience the same resistance in lbs. per ton and will burn the same coal per ton-mile ; but the H.P. per ton and the period of revolution of the screw will be proportional to the speed, and the steam pressure to the square of the speed.   L.

8. Taking the H.P. per ton as $\frac{1}{18}$ of the speed in knots, prove that this implies a resistance of about 20 lbs. per ton, or an equivalent incline of 1 in 112 ; and with a coal consumption of 2 lbs. per H.P. hour, the coal capacity required for a voyage of 3000 mls. is about a sixth of the tonnage. Calculate for a steamer of 26,000 tons at a speed of 23·5 knots.

L.

9. The mean density of the earth is 5·5 ; find the attraction between two small spheres a metre apart, each weighing one kilogram, if the earth's radius is $6·37 \times 10^8$ cms.

V. H.

10. If the force between two spheres, masses M_1 and M_2, d cms. apart, is GM_1M_2/d^2 where $G = 6\cdot66 \times 10^{-8}$, find the time of revolution of a dust particle round a metal sphere of density 10, when no other bodies act on the system and the dust particle just clears the sphere. L. H.

11. A particle moves uniformly round a circle with angular velocity ω, shew that its acceleration is towards the centre and equal to $\omega^2 r$. Assuming the Law of Gravitation, and taking the orbits of the Earth round the Sun and of the Moon round the Earth as circular, compare the masses of the Sun and of the Earth, given that the Moon makes 13 revolutions per year, and that the Sun is 390 times as distant as the Moon. L.

12. A train of weight 200 tons can be drawn by an engine at the uniform speed of 30 mls. an hr. up an incline of 1 in 200, or at 40 mls. an hr. up an incline of 1 in 400. Assuming the frictional resistance to be constant, calculate it in lbs. per ton; also the H.P. exerted by the engine, assuming it to be the same in both cases. L.

13. A stream of water flowing at the rate of 1100 lbs. per sec. turns an undershot water-wheel. It comes against the paddles with a velocity of 8 ft. per sec., and leaves with the velocity of the paddles, 4 ft. per sec. Find the rate of loss of momentum and the rate of working of the wheel.

L.

14. A bullet fired horizontally from a rifle pierces in succession three thin screens placed at intervals of 1000 ft.; and the times by chronometer of passing from the 1st screen to the 2nd, and from the 2nd to the 3rd, are found to be ·847 sec. and ·948 sec. respectively. Find the ratio of the retarding force (assumed to be uniform) to the weight of the bullet. L.

15. A hoop weighing 2 lbs. and 2 ft. in diameter is rolling at the rate of 30 revolutions per min. along a horizontal road. Find the kinetic energy of the hoop in ft. lbs. (The plane of the hoop is vertical, and the point of it in contact with the ground momentarily at rest.)   L.

16. A heavy particle falls down a rough inclined plane which makes an angle of 45° with the horizon; the velocity of the particle after falling down the plane is half the velocity which would have been gained by the particle if it had fallen freely through the vertical height of the plane. Find the coefficient of friction between the particle and the plane.

                                                                    L.

17. A ballistic pendulum whose time of complete vibration is 4 seconds, has a bob whose mass is 100 kilograms. A shot weighing 30 grams fired into the bob produces an oscillation of the pendulum of 1° on each side of the zero. Calculate the velocity of the shot.   V.

18. A solid cylinder 8 inches in diameter and weighing 5 lbs. rolls down a rough plane 10 ft. long and falling 1 in 16. Find the time taken, and the final angular velocity of the cylinder.   V.

19. How would you determine the moment of inertia of the pulley in an Atwood's machine? If the rim of this pulley were 10 cms. in diameter and the moment of inertia were 1500 c.g.s. units, what would be the acceleration of a system in which the weights are 200 and 250 gms. respectively at a place where $g = 981$ cms. per sec. per sec.?   V.

20. A wheel and axle of given dimensions in fixed bearings is started from rest by a weight attached to a flexible cord wrapped round the axle. Shew how to calculate the acceleration if the bearings be smooth, and how to estimate the friction if they are rough by an observation of the time taken for a given length to unwrap.   L.

21. A straight uniform rod of length a and mass m hinged at one end is released from a horizontal position; calculate the angular velocity of the rod when it passes through the vertical position.   V.

22. A thin uniform heavy rod of length l and mass m, pivoted freely at one end, is released from the horizontal position. If the kinetic energy is divided into two parts, one of rotation about the centre of gravity and the other of translation of the centre of gravity, calculate the amount of each type.   V.

23. A moveable mass is attached to a compound seconds pendulum; determine the positions in which the time will be (i) less than, (ii) equal to, and (iii) greater than, a second.
                 V.

24. A body performs torsional oscillations about a fixed axis; obtain an expression for its angular velocity in any position. Shew that its potential energy at any time is equal to $\dfrac{2I\pi^2a^2}{T^2}$ where I is the moment of inertia, a the angular displacement, and T the time of oscillation.   V.

25. A uniform wire of length $3a$ is bent into the form of an equilateral triangle. Calculate the length of the simple equivalent pendulum (i) for oscillations about a vertex in the plane of the triangle, and (ii) for similar oscillations in the perpendicular plane.   V.

26. Find the centre of oscillation of a square plate whose edge is a oscillating under gravity with its plane always vertical, when the point of suspension is (i) at a corner, (ii) at the middle of one side.   V.

27. A horizontal circular disc of radius a is suspended by a wire; calculate by how much the time of torsional oscillation will be altered, if the point of attachment, instead of being at the centre of the disc, lies at a distance $\frac{a}{10}$ from the centre, the disc being kept horizontal by a small weight attached to the rim equal to $\frac{1}{9}$ of the weight of the disc.   V.

28. A sphere of mass 98 gms. and radius 3 cms. is suspended by a wire of mass 2 gms. and length 50 cms. from a knife-edge of small mass. Find the length of the equivalent simple pendulum.   V.

29. A thin heavy hoop of diameter D is supported by a peg at its circumference and makes small oscillations in its own plane. Calculate the time of an oscillation.   V.

30. An inertia-bar, supported by a metal wire, vibrates torsionally. Compare the times of vibration of this system with a similar system in which (i) the linear dimensions of the inertia-bar are all doubled, (ii) the linear dimensions of the wire are all doubled, (iii) the linear dimensions of both bar and wire are all doubled.   V.

31. A vertical steel wire carries an inertia-bar at its lower end and the time of torsional vibration is 2·7 secs. If the linear dimensions of the wire are all doubled and those of the bar are halved, the materials being unchanged, find the new period of vibration.   V.

32. Given that Young's Modulus for steel is 2×10^{12} in c.G.s. units, calculate its value in pounds' weight per square inch, given that 1 ft. = 30 cms., and 1 kilogram = 2·2 lbs. (g = 981 cms. per sec. per sec.)   V.

33. A wire 10 metres long and 0·01 sq. cm. in cross-section is stretched 0·1 cm. by a load of 2·5 kilograms. Find the value of Young's Modulus in c.G.s. units.   L.

34. A horizontal beam, clamped at one end and free at the other, is loaded at the free end and the sag produced there is measured. How would the sag be affected by doubling the linear dimensions of the beam, the load remaining the same ?

V.

35. If a cylindrical beam 150 cms. long and of cross-sectional area 2 square cms. be loaded in the middle with 10 kilograms, calculate the sag if Young's Modulus for the material is 2×10^{12} c.g.s. units.

V.

36. A hollow metal tube 1 cm. in internal and 1·4 cm. in external diameter, whose length is 80 cms., is supported at the ends and loaded with 5 kilograms in the middle. Calculate the sag if Young's Modulus for the metal is 2×10^{12} c.g.s. units. ($g = 981$ cms. per sec. per sec.)

V.

37. A uniform wire 3 metres long and 2 mms. thick can be twisted 2·5 times round without permanent set. If the moment of the force it then exerts be equal to that of a weight of 200 gms. at an arm of 20 cms., calculate the average energy in ergs per c.cm. stored in the twisted wire.

V.

38. If a load of 2 kilograms at the end of a lever 30 cms. long, applied to one end of a circular rod $\frac{1}{3}$ cm. in diameter and $\frac{1}{2}$ m. long, twist its point of application through 10°, the other end being firmly clamped, what is the rigidity of the rod in c.g.s. units?

L. H.

39. Prove that $\dfrac{1}{Y} = \dfrac{1}{9k} + \dfrac{1}{3n}$, where Y, k, and n are respectively Young's Modulus and the bulk and rigidity moduli for a substance. If this equation is used to find k for a metal of which Y is about 2×10^{12} and n is about 10^{12}, these quantities being determined with errors of the order of one per cent., what may be the error in k?

L. H.

40. Compare the density of water at the surface and at the bottom of a lake 100 metres deep, given that the compressibility is $\frac{1}{22000}$ per atmosphere of 760 mms. of mercury, and that the specific gravity of mercury is 13·6.   L.

41. The barometer changes from 762 to 736·6 cms. between the two weighings of a drying-tube whose mass is 40 gms. The apparent increase in mass due to moisture absorbed is 0·5 gm. Find the error which would be produced if, in the correction for buoyancy, the change in the barometer were neglected; given that the density of air at the first observation is ·0012 gm. per c.cm., that the mean density of the drying tube and its contents is 2·4, and that the density of the weights is 8·4.   V.

42. One limb of a U-tube consists of a capillary of internal radius 0·05 cm., and the other of a similar capillary of radius 0·2 cm. If this U-tube be partially filled with a liquid of density 1·05 and surface-tension 50 dynes/cm., what will be the difference in the heights of the meniscus in each tube?
          V.

43. Given that the value of the surface-tension of water is 80 c.g.s. units, calculate its value in millimetre-milligram-minute units. A U-tube with vertical limbs is about half filled with water; calculate the difference in the height of the liquid in the two limbs if one limb is 1 cm. in diameter and the other 1 mm.   V.

44. A soap-bubble of 2 cms. diameter is situated within another of 3 cms. diameter. Find the diameter of a single bubble, the interior of which would be at the same pressure as the interior of the smaller bubble.   V.

45. If the surface-tension of water is 74 dynes per centimetre, what pressure will it produce in a bubble $\frac{1}{1000}$ cm. in diameter?   L. H.

46. A capillary tube is dipped vertically into mercury of surface-tension 540 C.G.S. units, and the surface in the tube (assumed to be part of a sphere) is found to be 1 cm. below the free surface outside. If $g = 981$ C.G.S. units, and the density of mercury is 13 6, calculate the radius of curvature of the meniscus.   V.

47. Express a surface-tension of 80 C.G.S. units in ft.-lb.-sec. units, given that 1 lb. = 453·6 gms.   V.

48. A capillary 0·1 mm. diameter has its lower end immersed in water, and the water rises in it to a vertical height of 30 cms. Calculate the surface-tension of the water. If the capillary tube were shorter than 30 cms. discuss what would happen.   V.

49. A drop of water is placed between two plane pieces of glass which are pushed close together until the film is everywhere x cms. thick and A sq. cms. in area. Find the force with which the plates cling to each other, the surface tension of water being T and its angle of contact with glass zero.   L.

50. What will be the pressure in a spherical cavity within a mass of water, if the cavity is ·001 cm. in radius and at a depth of 10 cms. below the surface of the water? The pressure of the air on the surface of the water is that due to 760 mms. of mercury, and the surface-tension of water is 78 C.G.S. units.   L.

51. Calculate the work done on the film in blowing a soap-bubble from a diameter of 3 cms. to one of 30 cms., if its surface-tension be 45 C.G.S. units.   L.

52. Prove that the pressure-excess inside a soap-bubble of radius R is $2gsh\dfrac{r}{R}$, where h is the height to which the soap solution of density s could rise in a tube of radius r. Work out the numerical value if the liquid is as dense as water, and if the diameter of the bubble equals the height the liquid could ascend in a tube $\frac{1}{4}$ mm. in bore. **L.**

53. A vertical glass tube is drawn out at its lower end into a capillary, and a mercury column extends from a point in the capillary to a point 37·5 cms. above it. Find the radius of the mercury meniscus in the capillary, if the surface tension of an air-mercury surface is 540 dynes per cm., and the density of mercury is 13·6. **V. H.**

54. A bubble of radius 10 cms. filled with air at 15° C. contracts till its radius becomes 5 cms. Find the mass of the air expelled, given that the mass of 1 c.cm. of air at N.T.P. is ·0013 gm., and that the surface-tension of soap solution in air is 70 C.G.S. units. **V. H.**

55. Prove that the saturated vapour pressure outside the surface of a sphere of liquid of radius a exceeds that outside a flat surface of the same liquid by $\dfrac{\rho}{\sigma - \rho}$ times the difference of pressure inside and outside the spherical surface ; σ and ρ being respectively the densities of the liquid and of the vapour surrounding it. (The sphere is to be supposed so large that the effect of curvature upon the vapour pressure is small compared with the vapour pressure itself.)

56. If the pressure at a certain level in a vapour obeying Boyle's Law be P, shew that the pressure P' at a level lower than the first by d is given by $P \log_e (P'/P) = g\rho d$, where ρ is the density of the vapour under pressure P. Shew that this equation may also be written $\log_e (P'/P) = gd/R\theta$, where R and θ have the same meanings as in the Gas Laws.

57. From the result of Q. 56 shew that the saturated vapour pressure P' outside a very small sphere of liquid of radius a is given by the equation

$$P \log_e (P'/P) = \frac{\rho}{\sigma} \left(P' - P + \frac{2T}{a} \right),$$

where T represents the surface-tension, P the vapour pressure outside a flat surface, and ρ the corresponding density of the saturated vapour. Shew that $P' - P$ may be neglected in comparison with $\frac{2T}{a}$; also deduce the result of Q. 55 from the foregoing equation.

58. Apply the approximate calculation of Q. 55 to find the value of $P' - P$ for water globules of radius (a) 1 mm., (b) ·001 mm., at temperatures 15° C. and 100° C. in each case. It is given that 1 litre of hydrogen at N.T.P. weighs ·09 gm., and the molecular weight of water is 18; the saturated vapour may be assumed to obey the gas laws; the density of water is 1 at 15° and 0·7 at 100°; the surface-tension is 75 dynes per cm. at 15°, and 60 at 100°; the vapour pressure of water at 15° and 100° is 12·67 mms. and 760 mms. of mercury respectively.

59. Work the previous question by the accurate formula of Q. 57, and thus shew that the result is scarcely affected. ($\log_{10} \epsilon = ·434$; a pressure of 760 mms. of mercury $= 10^6$ dynes per sq. cm.)

60. Assuming the surface-tension of water not to vary with the extent of surface, calculate by the methods of Q. 55 and Q. 57 the size of a globule outside which the saturated vapour pressure (a) at 15° C., (b) at 100°, is double the vapour pressure over a flat surface at the corresponding temperature. Use the data of previous questions.

61. If we have means at our disposal for just detecting variations of pressure equivalent to 0·1 mm. of mercury, calculate by the formula of Q. 55 the radius of the coarsest capillary tube within which the effect of curvature on the saturated vapour pressure of water will be noticeable at 100° C., assuming the surface-tension of water to be 60 dynes per cm.

62. Calculate the height to which the water would rise in the capillary in the previous question.

63. A bicycle tyre leaks at the valve where it is found that if wetted a bubble one-quarter of an inch in diameter is blown in from 55 to 65 seconds. How long will it take for the pressure to fall to half its amount? The inner diameter of the air-tube may be taken as $1\frac{1}{4}$ inch and the diameter of the axis of the air-tube as $26\frac{1}{2}$ inches.   L. H.

SOUND.

64. If an engine whistle appears to change in frequency in the ratio 9 : 8 in passing a hearer, find the ratio of the velocity of the engine to the velocity of sound.   L.

65. An engine in a cutting between two bridges is whistling when its velocity is $\frac{1}{20}$ that of sound in air. Compare the frequencies of the echoes from the two bridges to an observer between them.   L.

66. A train is approaching a station at the rate of 60 miles per hour, and a whistle giving 1000 vibrations per second is sounded on the train.   What will be its apparent frequency to an observer at the station ?   If a similar whistle is sounded at the station, what will be its apparent frequency to an observer on the train ?   The velocity of sound is 1100 feet per second.   V.

67. A bar of density 8 and length 1 metre vibrating longitudinally gives a fundamental three octaves above middle C (256); find Young's Modulus for it.   L.

68. What is the elasticity of a rod a metre long and seven times as heavy as water, which vibrates longitudinally 1200 times a second?   V.

69. Calculate Young's Modulus for a given sample of wood from the observations that a yard long of it floats vertically in water with two inches protruding, and emits when stroked longitudinally a note three octaves above the middle C (256).   L.

70.   A bar 2 metres long having a section 1 cm. square is supported by its ends and droops 2 cms. for an extra load of 1 kilogram at the centre. Calculate Young's Modulus. If the bar weighs 600 gms., what is the velocity of sound along it ?   V. H.

71.   A rod of specific gravity 0·7 is found to be stretched one millimetre when suspended vertically and loaded with a weight of lead of specific gravity 11, the volume of the weight being equal to that of the rod. What note will the rod alone emit when held in the middle and rubbed longitudinally ?   V.

72.   What is the frequency of the gravest note which can be sounded by a steel wire 2 mms. in diameter and 150 cms. long when stretched by a weight of 50 kilograms ? The specific gravity of steel is 7·7.   V.

73.   A string is under such tension that its length is increased by $\frac{1}{n}$ of its original value. Shew that the frequency of the fundamental longitudinal vibration is $\sqrt{n}$ times the frequency of the fundamental transverse vibration.

L. H.

74.   One end of a flexible cord, whose mass is 0·1 gm. and length 392 cms., is attached to the end of a prong of a large tuning-fork, and hangs vertically with a weight of 36 gms. at the lower end. When the prong is horizontal and vibrating in a vertical plane the cord breaks up into four segments ; calculate the pitch of the fork.   V.

75.   What is the velocity of sound in a gas, in which two waves of length 1 and 1·01 metres respectively produce 10 beats in 3 seconds ?   V.

76. Joly's value for the specific heat of air at constant volume is $0.17151 + 0.02788s$, where s is the density. The specific heat at constant pressure is 0.2389; find the velocity of sound in air of density 0.001293, the barometer standing at 760 mms. of mercury of density 13.6, and the value of g being 981 cms. per sec. per sec.   L.

77. A series of closed organ pipes are arranged in order of gradually diminishing length. Any one of the pipes gives seven beats a second with the one next to it. Two beats are heard when the 35th and 1st are sounded together, and five when the 36th and 1st are sounded. What is the length of the longest pipe, if the velocity of sound in air is 340 metres per sec. ?   V. H.

78. The velocity of sound in water is 1440 metres per second; calculate the diminution in volume of a litre of water when subjected to an extra pressure of 10^6 dynes per sq. cm.   V.

79. Two tuning-forks vibrating 200 times per second are in perfect unison when at the same temperature, but when there is a difference of 10° the Lissajous curves formed by them are found to pass through a complete cycle in 5 seconds, the warmer fork vibrating more slowly. Find the temperature coefficient of the frequency of the fork.   V.

80. A tuning-fork gives a particular note at 15° C. It is put in boiling water, and immediately after it is taken out its frequency is lowered one per cent. The coefficient of linear expansion of the metal is 12×10^{-6}. What is the temperature-change in the elastic constant?   L. H.

81. Given log 2 and log 3, calculate the intervals between true and tempered fifths, and between true and tempered fourths.   L. H.

HEAT.

82. A barometer with a brass scale reads 75·80 cms. at a temperature of 16° C. What will be the reading reduced to 0° C., if the coefficient of linear expansion of brass is 18×10^{-6} and that of cubical expansion of mercury is 18×10^{-5}?
                                                                V.

83. The metre being defined as the distance between two parallel lines on a platinum bar at 0° C., and the yard as the distance between two similar lines on a brass bar at 18° C., express the metre in terms of the yard, given the result of a direct comparison of the two at 18° C. as 1 metre = 1·09393 yds.; coefficient of expansion of platinum = 86×10^{-7}.   V.

84. If a crystal have a coefficient of expansion 13×10^{-7} in one direction and of 231×10^{-7} in every direction at right angles to the first, calculate its coefficient of cubical expansion.
                                                                L.

85. A crystal having principal coefficients of expansion ·000010, ·000012, and ·000015, weighs 50 gms. in vacuo and 30 gms. in water at 4° C. Find its weight in water at 20° C. of density 0·99827.   V. H.

86. Shew that in the case of a water-in-glass dilatometer the temperature of apparent maximum density will be above the true temperature, and find how much mercury should be placed in the bulb of such a dilatometer in order that the

temperature of apparent maximum density may be identical with the true temperature, given that the coefficients of dilatation of mercury and glass near 0° C. are ·000180 and ·000023 respectively.　　　　　　　　　　　　　V. H.

87. Explain how the force that can be exerted by a bar when cooling is connected with its coefficient of linear expansion for temperature and its coefficient of elongation for traction.　　　　　　　　　　　　　　　　V.

88. A lump of quartz which has been fused is suspended from a "quartz fibre" and allowed to oscillate under the influence of the torsion of the fibre. If the coefficient of expansion of the material is 0·0000007 and the temperature-coefficient of its rigidity is + 0·00013, how many seconds a day, or what fraction of a second a day, would a change of temperature of 1° C. make?　　　　　　　　L.

89. A watch has a non-compensated brass balance-wheel and a steel balance-spring. How many seconds will it lose per day for 1° rise of temperature, if the coefficient of linear expansion of brass is $18 \div 10^6$, and if Young's Modulus for steel at $t°$ is $2 \times 10^{12} \left(1 - \dfrac{2t}{10^4}\right)$?　　　　　　　L.

90. Explain the cause of draught in chimneys. What average excess in temperature in a chimney 136 feet high would under ordinary circumstances produce a pull measured by 1 inch on the water gauge?　　　　　　　L. H.

91. Describe Bunsen's calorimeter. Given that 1 c.cm. of ice at 0° C. yields ·9178 c.cm. of water at 0° C., determine the specific heat of a body whose mass is 1 gm. and temperature 100° C. which when dropped into the instrument produces a shrinkage of 6 cubic millimetres. (Latent heat of fusion = 80.)　　　　　　　V.

A.　　　　　　　　　　　　　　　　　　　　　2

92. Determine the ratio of the specific heats of air from the following data. Velocity of sound = 34215 cms. per sec. in air at 750 mms. and 17° C.; density of air = ·00129 gm. per c.cm. at N.T.P.; coefficient of expansion of air = $\frac{1}{273}$; $g = 981$ cms. per sec. per sec.; density of mercury = 13·6 gms. per c.cm.   V.

93. Calculate the specific heat of a gas from the following data. The gas passed through the calorimeter was supplied by a vessel of capacity 30 litres. At the beginning of the experiment the pressure in the vessel was 7 atmospheres, while at the end it had fallen to 3 atmos. The temperature of the gas was such that under atmospheric pressure its density would have been ·0012. The difference in the temperature of the gas on entering and leaving the calorimeter was 208°, and the corrected rise of the temperature of the calorimeter was 10°·2. The water equivalent of the calorimeter and contents was 700 gms.   V.

94. The thermal conductivity of a substance is represented by 1 in c.G.s. units; what will be its numerical value in millimetre-milligram-minute units?   V.

95. A tin cylinder 40 cms. in diameter and 50 cms. long is covered all over by a layer of felt 0·33 cm. thick. Steam at 100° C. is passed through the cylinder, the external temperature being 20° C., and water is found to accumulate at the rate of 3 grams per min. If the Latent Heat of steam at 100° is 537, find the conductivity of the felt.   L.

96. A small quantity of heat is to be measured (i) by Bunsen's ice calorimeter, (ii) by Favre and Silbermann's method, in which the heat is imparted to the mercury in a large thermometer bulb, (iii) by the same method with ether in place of mercury. Compare the sensitiveness of the three ways. Latent heat of fusion of ice = 80 ; density of ice = 0·92 ;

specific heat of mercury $= \frac{1}{30}$, of ether $= \frac{1}{2}$; coefficient of expansion of mercury $= ·00018$, of ether $= ·00165$; density of mercury $= 13·6$, of ether $= 0·7$.

97. Assuming Newton's Law of Cooling, compare the rates of loss of heat in vacuo of two copper spheres whose radii are 10 cms. and 5 cms., their temperatures being respectively 20° and 10° above that of the enclosure in which they are placed. Compare also the rates of fall of temperature. (Both spheres conduct heat perfectly.)

98. Give the relative rates of cooling of two similar bodies at 327° and 127° C. respectively, the surrounding bodies being all at 27°, (i) on Newton's law of cooling, (ii) on the supposition that radiation is proportional to θ^4 and that convection may be neglected. V.

99. Calculate from the following data what would have been the final temperature of a calorimeter and its contents after the introduction of a hot body, had all loss by radiation been avoided. Initial temp. of air and water $= 15°·3$ C.; after introducing the hot body, the temp. at the end of successive minutes was 18·8, 20·0, 20·6, 20·7, 20·7, 20·6, 20·5, 20·4, 20·3, 20·2, 20·1, 20·0. V. H.

100. Find the weight of a litre of moist air at 20° C., the dew-point being 11° and the barometer standing at 76 cms. (Vapour pressure of water at 11° $= 1$ cm. approx.; density of dry air at 76 cms. pressure and 20° C. $= ·00120$; relative density of aqueous vapour and air $= 0·624$.) V.

101. Calculate the dew-point for air $\frac{3}{4}$ saturated at 18° C., given that for pressures of 6, 9, 12, 15 mms. of mercury respectively, the corresponding boiling-points of water are 4°, 10°, 14°, and 18° C. V.

102. A mixture of air and of the vapour of a liquid in contact with excess of the liquid is contained in a vessel of constant volume. At a temperature of 15° C. the pressure in the vessel is 70 cms. of mercury, at 30°, 45°, and 60° the pressures are 88, 110, and 145 cms. respectively. If at 15° C. the vapour pressure of the liquid is 15·4 cms., what is it at 30°, 45°, and 60° ?   L.

103.   Assuming Boyle's Law to hold, find the weight of dry air in a vessel containing 300 c.cms. of air saturated with aqueous vapour at 20° C. and subject to a total pressure of 73·74 cms. of mercury ; the density of air at N.T.P. being ·001293 gm. per c.cm. and the maximum pressure of aqueous vapour at 20° being 1·74 cm. of mercury.   V.

104.   Calculate the average molecular velocity in the case of a gas whose density at a pressure of 50 cms. of mercury is 0·001 gm. per c.cm.   V.

105.   Calculate the molecular velocity (of mean square) in the case of a gas whose density is 1·4 gm. per litre at a pressure of 76 cms. of mercury.   Density of mercury $= 13·6$; $g = 981$ cms. per sec. per sec.   V.

106.   Shew how the mean free molecular path in a gas may be obtained when the viscosity is known, and find the mean free path in air regarded as a uniform gas, given that the pressure is 10^6 dynes per sq. cm., the density 12×10^{-4} gm. per c.cm. and the viscosity 17×10^{-5} C.G.S. unit.

L. H.

107.   If a gas is suddenly compressed to $\frac{1}{10}$th of its original volume, obtain an expression from which the resulting temperature could be calculated.   V.

108.   (i) Air at 15° C. is suddenly compressed to $\frac{1}{10}$th of its volume ; find the temperature to which it will rise. (ii) If the compression raises the pressure to 10 times its original value, find the final temperature.   L.

109. What pressure would be required to compress dry air at the standard pressure and temperature into $\frac{1}{30}$ of its volume (*a*) slowly, (*b*) suddenly; and what rise of temperature would result? (Ratio of specific heats, 1·4.) V.

110. If ordinary dry air at ten atmospheres and 15° C. is suddenly released, find with the help of logarithms the new temperature. V.

111. It was found by Noble and Abel that the explosion of 1 gm. of gunpowder in a closed space of 1 c.cm. exerted a pressure of 6300 atmospheres and left a liquid residue of 0·6 c.cm.; while the gaseous products drawn off to 0° C. and 760 mms. occupied 280 c.cms. Deduce from these data the temperature attained in the explosion. L. H.

112. Shew how to calculate the difference between the two specific heats of a perfect gas, given that a gram of it occupies 1000 c.cms. at a temperature of 27° C. and one mega-dyne per sq. cm. pressure. ($J = 4\cdot2 \times 10^7$ ergs per calorie.) V.

113. Calculate the intrinsic energy of air at 15° C. and 750 mms. pressure, assuming the ordinary values of such physical constants as you require.

114. From what height must a lump of ice at 0° C. fall in order to melt itself, supposing it possible to concentrate all the energy of the fall in the lump? L.

115. Find the velocity of a lead bullet which warms itself 250° C. by striking an unyielding target, the whole heat generated by the impact heating the lead. (Specific heat of lead = ·033.) V.

116. Two bars of iron, each weighing 50 kilograms, are supported horizontally, each by two vertical cords 100 cms.

long, all four strings being in the same vertical plane, in such
a way that the axes of the bars are in the same horizontal
line, and their nearer ends a short distance apart. In the
space between the bars and just touching each is placed
1 kilogram of lead of specific heat $\frac{1}{30}$. One of the bars is
pulled out in the plane of the cords till they are inclined at
60° to the vertical, and then the bar is released so as to
impinge directly on the lead and crush it. After the impact
the horizontal amplitude of oscillation of the whole system is
found to be 20 cms. on each side of the position of equilibrium.
Calculate the rise in temperature of the lead, supposing all
the heat produced by the impact to be confined to it.

V. H.

117. Calculate the internal and external work done in
vaporising a gram of water at 100° C., given that the latent
heat of evaporation is 536 ; the value of Joule's equivalent is
42×10^6 ergs per calorie, the volume of a gram of steam at
100° C. and 760 mms. pressure may be taken as calculable by
the gas laws from its molecular weight, the specific gravity of
mercury is 13·6, and the value of g is 981 cms. per sec.
per sec.

118. The latent heat of steam at 100° C. is 536; calculate
what fraction of this heat is used up in performing external
work during vaporisation, assuming the density of steam at
100° C. to be ·0007, and an atmosphere to be 10^6 dynes per
sq. cm.   V.

119. Assuming that steam obeys the gas laws, shew that
the work done in changing the volume of 1 gm. of steam at
100° C. and 760 mms. to the volume at 101° C. and 787 mms.
(the saturation pressure at 101° C.) is more than enough to
supply the heat needed for the rise in temperature ; the
specific volume of steam at 100° being 1700, and its specific
heat at constant pressure 0·48.   L. H.

120. Calculate the change of entropy when a gram of water at 30° C. is mixed with 2 gms. at 0°, assuming that the specific heat of water is unity at all temperatures.

121. Calculate the change of entropy when a gram of steam at 100° C. cools to water at 0°, assuming that the latent heat of vaporisation is 536 and that the specific heat of water is 1 at all temperatures.

122. An engine of eight H.P., supposed reversible, works between the temperatures 100° C. and 15° C. How much water should be evaporated in the boiler per hour, given the appropriate values of Joule's equivalent and the latent heat of steam?

123. Coal is burnt at a temperature of 1527° C., and the heat is used (a) to warm the air of a room at 7° C., (b) to drive a reversible heat engine whose condenser is at the temperature of the room. In case (b) the heat engine drives backwards another reversible heat engine working between temperatures 7° C. and 27° C., and the heat escaping from the latter is used to warm the air of the room. Compare the two processes (a) and (b) from the point of view of economy.

124. Calculate the volume of a gram of steam at 100° C., given that the latent heat of evaporation of water at 100° C. = 535, and the change of boiling point is 0°·37 C. per cm. of mercury pressure. $(J = 4·2 \times 10^7.)$ V. H.

125. Calculate the change in the boiling point of water for a fall of the barometer from 76 to 75 cms., assuming that a gram of steam occupies 1700 c.cms., and taking the usual values for other constants.   V. H.

126. If sulphur has a density 2·05 just before, and 1·95 just after melting, the melting point being 115° C. and the latent heat 9·3, find the alteration in melting point per atmosphere change of pressure.   L. H.

127. A spring when pulled out at 15° C. is found to require to be supplied with heat equivalent to $\frac{1}{10}$ of the work done in its extension to prevent its temperature from falling. Find the temperature-coefficient of pull for a given extension.

L. H.

128. The surface tension of a water-air surface is **77** dynes per cm. at 0° C. and 60 dynes per cm. at 100° C. Compare the amount of work required to blow a bubble of given size isothermally and adiabatically.   V. H.

LIGHT.

129. The mean interval between successive eclipses of Jupiter's second satellite is 42 h. 28 m. 42 s. Calculate the greatest and least intervals, given that the velocity of light is 3×10^{10} cms. per sec., and that the earth's mean orbital velocity $= 3 \times 10^6$ cms. per sec. V. H.

130. Assuming the satellite of a planet to revolve once in 48 hours, calculate the greatest and least interval between two successive eclipses of the planet as observed from the earth. Distance of sun from earth $= 150 \times 10^6$ kilometres; velocity of light $= 3 \times 10^5$ kilometres per sec. V. H.

131. The wave-length of a line in the solar spectrum is found to be 6×10^{-5} cm. at the sun's pole, and is changed by 4×10^{-10} cm. at that part of the sun's equator which is moving in the line of sight. Calculate the linear velocity at the equator, if the velocity of light is 3×10^{10} cms. per sec.

V.

132. In the spectrum of a star the more refrangible component of the double sodium line is found to coincide in position with the less refrangible component of the pair in the solar spectrum. Calculate the velocity of the star, and state whether it is moving towards the earth or away from it. (The wave-lengths of the two sodium lines are 5889 and 5895 "tenth metres" respectively.) V. H.

133. A concave spherical reflector is placed with its axis vertical so as to form a shallow tray. The position of the centre of curvature is 100 cms. above the surface of the tray. Calculate the position of the apparent centre of curvature, *i.e.* that point on the axis for which an object and its image are coincident, when the tray is filled with water of index $\frac{4}{3}$.

V.

134. You are given an achromatic objective of 12 inches focal length, and are asked to construct an inverting telescope focussed for infinity and magnifying twelve times. State what lenses you would require for the eye-piece. V.

135. Two telescopes placed one behind the other magnify 10 and 20 times respectively ; each has been so adjusted that the rays from an object at infinity emerge parallel. Prove that the optical system so obtained will be a telescope, and find its magnifying power (*a*) when the eye-pieces are together, (*b*) when the objectives are together, (*c*) when an eye-piece and an objective are together. V.

136. A telescope with aperture 4 inches is provided with two eye-pieces magnifying 10 times and 100 times respectively. It is used to view (i) a star, (ii) a luminous surface, such as the sky by day. Supposing the eye to have an effective aperture of $\frac{1}{6}$ inch, what will be the ratio of the brightnesses (i) of the star, (ii) of the sky, when seen by the naked eye and with the telescope, with each of the eye-pieces ? Which eye-piece would be best for seeing a star by day ?

I? H.

137. If an achromatic converging lens has to be constructed of two materials, one of which has refractive indices 1·36 and 1·38 for red and blue rays, while the corresponding indices of the other are 1·78 and 1·80, which material should be used for the concave lens of the combination ? If the compound lens is to have a focal length of 50 cms., what should be the focal lengths of the two constituents?

V.

138. Calculate the focal lengths of the separate lenses in an achromatic combination for which the focal length is to be 30 inches from the following spectroscopic measurements :— crown glass, $\mu_c = 1\cdot516$, $\mu_r = 1\cdot525$; flint glass, $\mu_c = 1\cdot619$, $\mu_r = 1\cdot636$.

V.

139. Find the power of the single lens equivalent to two thin coaxial lenses of powers p_1 and p_2 separated by an interval d.

L. H.

140. A person whose range of distinct vision is between 10 and 20 cms. respectively puts on spectacles which remove the near point to 20 cms. ; what is the focal length of the spectacle lens, and what is the greatest distance to which he will see distinctly with the help of the spectacles?   V.

141. A double convex lens whose surfaces are equally curved is 3 inches in diameter and $0\cdot1$ inch thicker at the centre than at the circumference. Its refractive index is $1\cdot6$; find its focal length.

V.

142. A converging lens has one face flat and the radius of the other is 2 ins. ; the thickness of the lens is $1\cdot2$ ins. It is found that an object 8 ins. away from the convex face forms an image of the same size at $7\cdot2$ ins. from the flat face. Calculate the refractive index of the glass.

143. The lens of Q. 142 is used to form an image of a small object on the axis at $11\cdot2$ ins. from the flat face ; calculate the distance of the image from the curved face.

144. Parallel light is incident on the convex side of a plano-convex lens 1 in. thick and 3 ins. in diameter, its edges being sharp and its refractive index equal to $1\cdot5$; find how far from the flat face the light will be brought to a focus.

145. Calculate the position of the principal points of a double convex lens whose radii of curvature are 20 and 30 cms. respectively, the glass being 1 cm. thick and having a refractive index of 1·5.   V. H.

146. Three convex lenses are each 1 in. thick and 3 ins. in diameter with sharp edges, and are made of glass of refractive index 1·5. The first has one flat face, the second has faces of equal curvature, the third has a concave face which exactly fits the faces of the second. Calculate for each lens the focal length and the distance of the principal foci from the corresponding faces.

147. Determine the focal length, radii of curvature, and refractive index of a biconvex lens 1 inch thick from the following axial observations. An object 6 ins. from the nearer face A gives a real image 8 ins. from the other face B; a bright point at $1\frac{1}{7}$ in. from B and $2\frac{1}{7}$ ins. from A gives an image by reflection coinciding with itself; the same thing happens if the bright point is $1\frac{1}{2}$ in. from A and $2\frac{1}{2}$ ins. from B.

148. The lens of Q. 144 has air on one side and water (refractive index $\frac{4}{3}$) on the other. Determine the position of the nodal points according as the water adjoins (i) the flat or (ii) the curved face ; also of the focal points.

149. An object is situated on the axis of the lens of Q. 148 at a distance of 6 ins. from the nearer face. Determine the distance of the image from the nearer face in each of the possible cases ; also find the magnification.

150. If in a normal eye the distance of the " second nodal point" from the retina is 15 mms., what will be the area covered on the retina by the image of a circular disc 30 cms. in diameter, the centre of which is 2 metres distant from the eye in the direct axis of vision and whose plane is at right angles to this axis ?   L.

151. An equiconvex glass lens (refractive index $\frac{3}{2}$) with sharp edges is 1 in. thick and 3 ins. in diameter. It separates air on one side from water (refractive index $\frac{4}{3}$) on the other. Determine the position of the nodal points and focal points.

152. An object is placed on the axis 6 ins. from the nearer face of the lens of Q. 151, (i) on the side where there is air, (ii) in the water. Find the position of the image in each case, and also the magnification.

153. A zone-plate exposed to parallel rays of light incident normally upon it brings the light to a focus 5 inches away; if the wave-length of the light is ·00002 inch, what is the outer radius of the first opaque zone, the centre being transparent? If the plate receives light from a point on its axis 15 ins. away, where will be the conjugate focus?

154. A zone-plate intended for use with the lantern has the outer radius of the first opaque zone equal to 2 mms.; how far off will be its principal focus for light of wave-length ·00005 cm.? (The central area is transparent.)

155. The object-glass of a telescope is covered by a screen containing two narrow parallel apertures 12 cms. apart, and an illuminated slit placed parallel to the apertures is looked at through the telescope. If the light from the slit is of wave-length 5×10^{-5} cm. and the telescope magnifies 20 times, calculate the angular distance between the two central dark bands. V.

156. Describe Fresnel's biprism experiment. The wave-length of sodium light is 5892×10^{-8} cm.; find the distance between the interference bands supposing the distances between slit and biprism, and between biprism and screen, to be each 50 cms. The angle of the biprism is 179° and its index of refraction is 1·5. V.

157. Describe in detail how you would arrange an optical bench so as to exhibit interference fringes with a biprism. How would you measure the distance between the two virtual images of the slit formed by the biprism ? If this distance is 0·2 cm., the wave-length of the light is 5900 tenth-metres, and the screen is 100 cms. from the slit, calculate the distance between two consecutive fringes. V.

158. What relation exists between the diameters of successive Newton's rings when a convex lens in optical contact with a flat surface is viewed by monochromatic light ? If two such successive rings have diameters of 3 mms. and 3·1 mms. when viewed normally, calculate the radius of curvature of the convex surface. V.

159. The convex face of a lens rests on a flat plate, and the Newton's rings formed are viewed normally by reflected light from a sodium flame. The wave-length is 5897 tenth-metres, and the diameters of two consecutive rings are 1 cm. and 1·01 cm. ; calculate the radius of curvature of the lens.

160. Calculate approximately (neglecting multiple reflections) the thickness of an air film which obliterates light of wave-length 5×10^{-5} cm. from the reflected beam, the rays in the film passing at an angle of 60° to the normal.

V. H.

161. Calculate the thickness of an air film between plates of a substance having a refractive index of $\sqrt{2}$, if a wave-length of 5×10^{-5} cm. as measured in air is absent in the reflected beam, the angle at which the ray strikes the film being 30°. V.

162. Calculate the thickness of a soap-film (refractive index $\frac{4}{3}$), which, when white light is incident on it at 45°, gives a dark band in the reflected light whose centre is found by the spectroscope to correspond to wave-length 6×10^{-5} cm.

163. The thickness of a certain object is known to be between 1·015 and 1·020 mm. When it is expressed in terms of four standard wave-lengths (6438·9, 5086·3, 4800·0, and 4678·9 "tenth-metres") the fractions of a wave are found by experiment to be ·61, ·14, ·55, ·51 respectively. Shew that (within the possible limits) these fractions agree best with the integers 1581, 2002, 2121, 2176 respectively; and hence calculate the thickness accurately. V. II.

164. A beam of plane polarised light is changed into circularly polarised light by passing through a slice of crystal 0·003 cm. thick; calculate the difference in the refractive index of the two rays in the crystal, assuming this to be the minimum thickness that will produce the effect, and that the wave-length is 58×10^{-6} cm. in air. L.

MAGNETISM.

165. The value of H at Kew is about $0\cdot18$ c.g.s. unit; express this in mm.-mgm.-min. units.

166. At what points do the lines of force due to an ideal magnet run at right angles to its magnetic axis?

167. A magnet of moment M is placed (i) East and West, (ii) North and South; find the points in the same horizontal plane as the magnet at which the earth's horizontal field is just annulled.

168. A solid cylindrical bar of magnetised steel, 10 cms. long and 1 cm. in diameter, vibrates once in 10 secs. due to the earth's horizontal field when suspended so as to oscillate about an axis perpendicular to its length. If the density of steel is 8, and $H = 0\cdot18$, calculate the value of the magnetic moment.   V.

169. A weight of $0\cdot1$ gram, acting at 7 cms. from the centre of gravity of a dip magnet, causes the magnet to set horizontally. The same weighted magnet oscillating in a horizontal plane has a time of oscillation of $6\cdot28$ seconds. Calculate to one per cent. the tangent of the angle of dip, if the moment of inertia of the combined system is 700 c.g.s. units.   V.

170. A small magnet placed end-on at a metre from a declination magnetometer deflects it through 1° 30′

(tan 1° 30' = ·0262) ; and when placed end-on at 2 metres from a bifilar magnetometer deflects it through 7 divisions of its scale. Calculate the percentage of change in the horizontal component of the earth's magnetic intensity that is represented by each division of the scale of the bifilar, assuming the average value to be $H = ·1811$.   L.

171. Find an expression for the couple which a small magnet B exerts on a second small magnet A, whose axis produced bisects that of B at right angles. Find also the couple exerted by A upon B, and reconcile the result with the axiom that "action and reaction are equal and opposite."   V.

172. Two magnets of the same steel, dimensions $10 \times 3 \times ·5$ and $20 \times 4 \times ·7$ respectively, were found, when pivoted horizontally in the earth's field so as to oscillate about their shortest axes, to swing at the same rate. Compare the intensities of magnetisation of the two.   V.

173. Two short magnets, each of moment M, are so placed as to make angles of 45° with the line joining their centres. Find the strength and direction of the magnetic field at the point midway between them in each of the possible cases.

174. If two short magnets of equal moment are placed with the line joining their centres along the axis of one and perpendicular to the axis of the other, calculate the intensity and direction of the field at the middle point of the joining line.   V.

175. Two equal linear magnets are placed so that their directions are at right angles to each other. Find the direction and intensity of field at the point towards which both magnets are directed. (Avoid choosing a special case.)

                                                 V.

A.   3

176. Three short magnets, each of moment M, are placed at the corners of an equilateral triangle whose side is a, two lying along a side of the triangle with the North poles towards each other, while the third is perpendicular to that side. Find the strength and direction of the magnetic field at the centre of the triangle in each of the possible cases.

177. Two short magnets, each of moment M, occupy two corners of an equilateral triangle, and lie along the side joining them, its length being a. Find the strength and direction of the magnetic field in each possible case, (i) at the third corner, (ii) at the centre of the triangle.

178. The magnets in Q. 177 are turned so that each lies along the side joining it to the third corner. Find the strength and direction of the magnetic field there and at the centre of the triangle in each possible case.

ELECTRICITY.

179. Two small balls each weighing 0·1 gm. are suspended by silk fibres 100 cms. long from the same point. A charge of electricity is now communicated to them so that they separate till the distance between their centres is 4 cms. Calculate the charge on each, assuming the electricity to be uniformly distributed over the surfaces. V.

180. Calculate the capacity of a pair of large parallel conducting plates of area A when the distance between them is small and represented by d. Shew further that the field between the plates remote from the edges is uniform, and find its value for unit potential difference. V.

181. A thunder-cloud is over still water. What is the electrostatic surface-density of the charge on the water if under the cloud it rises 0·1 cm. above its previous level?
L.

182. If the area of each of the plates of a parallel plate condenser is 1000 sq. cms., and if their potential difference is 100 volts with an air-space of 0·1 cm. between them, calculate in ergs the energy stored in the condenser. (Velocity of light $= 3 \times 10^{10}$ cms. per sec.) V.

183. Calculate to one per cent. the force in gms.' weight exerted by one of the two plates of a condenser on the other,

3—2

the plates being at a distance of 0·01 cm., the area of each plate being 100 sq. cms., and the difference of potential being 1000 volts. The effect of the edges may be neglected.

V.

184. What potential would enable the tension on an insulated sphere of 3 cms. radius to balance $\frac{1}{2000}$ of an atmosphere? L.

185. A sphere 24 cms. in diameter has a charge of 15 units. When connected to a quadrant electrometer it gives a deflection of 48 divisions. It is then joined by a fine wire to a sphere of 12 cms. diameter, and the deflection is now only 38. Find the capacity of the electrometer and the charge on the small sphere. L.

186. A charged particle is placed near a conducting spherical surface. Shew that even though the sphere and the particle are charged with electricity of the same sign, the particle will be attracted to the sphere unless its distance from the surface of the sphere exceeds a certain value. Find an expression for this value. L. H.

187. A microfarad condenser has its plates charged and is then allowed to leak through a very high resistance R. If the potential difference falls to half its original value in 1 minute, find the value of R, given that $\log_e 2 = 0·69$.

V. H.

188. What is the value (a) in electromagnetic c.g.s. units, (b) in coulombs, (c) in electromagnetic mm.-mgm.-min. units, of a quantity of electricity whose electrostatic measure is 250 c.g.s. units? L. H.

189. Calculate in terms of a microfarad the capacity of an air condenser in which the plates are parallel and each a sq. metre in area, the distance between them being 1 millimetre. V.

190. A condenser of capacity 1 microfarad is charged by means of a Daniell cell of E.M.F. $1\frac{1}{2}$ volt. If the charge given to it were used to raise metal spheres of 5 cms. radius to a potential of 20,000 volts, how many such spheres could be charged?

191. An electric motor developes 5 horse-power in its shaft with a current of 45 amperes at 100 volts. What is the efficiency of the motor, a horse-power being taken as equivalent to 746 watts?   V.

192. Electrical energy is sold at the rate of 4d. per kilo-watt-hour. The mechanical equivalent of the heat given by the burning of coal worth 4d. is 10^8 foot-lbs. Compare the prices of the two forms of energy. Why is electrical energy so much dearer than coal energy? (1 H.P. = 746 watts.)

L.

193. How many ergs correspond to 1000 watt-hours? A lamp of 400 candle-power requires $2\frac{1}{2}$ watts per candle; how many amperes will it take at 100 volts? What will be the loss of E.M.F. on passing the current through a conductor whose resistance is 0·2 ohm?   V.

194. (i) Calculate the value of the energy stored in a condenser formed of two parallel plates 1 metre square separated by 1 mm. of air, the potential difference being 100 volts; the cost of a Board of Trade unit (1000 watt-hours) is 4d. (ii) A condenser built up in this way is to store a pennyworth of energy; calculate the combined thickness of all the air-gaps.

195. Twenty-one calories are produced in 5 minutes in a wire by a current of one-sixth of an ampere. Calculate the number of watts required to keep up this current and the difference of potential at the ends of the wire. (1 calorie = 4·2 joules.)   V.

196. Calculate the power required to light 80 incandescent lamps, if the E.M.F. required be 65 volts and the current required by each be 0·8 ampere. If the lamps be all in parallel and the leads have a resistance of 0·5 ohm, calculate the power wasted in them.   L.

197. The resistance of a copper wire and a column of electrolyte in series is 100 ohms at 0° C. and 90 ohms at 20° C. If the temperature coefficient of resistivity of copper is ·004, and that of conductivity of the electrolyte is ·06, find the resistance of the copper wire.   V. H.

198. A current is passed between two plates immersed in a solution of copper sulphate and measured by a tangent galvanometer. The deflexion of the galvanometer is 45°, and after an hour it is found that 216 mgms. of copper have been deposited. Being given that unit current deposits 19·6 mgms. per minute, find what current would cause a deflexion of 60°.
                                                                    V.

199. If the chemical action going on in a battery results in the formation of 500 calories of heat for every gram of metal dissolved, and if the quantity of electricity set in motion by the solution of a centigram of metal is 0·5 c.G.s. unit, what is the E.M.F. of the battery (assumed to be calculable from the above data)?   V.

200. The electro-chemical equivalent of zinc is 0·00034 gram per ampere-second. Find the cost of the zinc used in a primary battery for each H.P. per hr., if zinc costs p pence per kilogram, and if the cell in which the zinc is used gives V volts.   (1 H.P. = 746 watts.)   L.

201. A normal Daniell's cell has an E.M.F. of 1·07 volt and resistance 2 ohms. Its terminals are connected by two wires in parallel of 3 and 4 ohms. Assuming that the electro-chemical equivalent of copper is ·000328 gram per coulomb, calculate the weight deposited in the cell, and also the heat developed (a) in the cell, (b) in each of the wires, during one hour of working of the cell. V.

202. The heat of combustion of hydrogen and oxygen to water is 34200 water gram units for each gram of hydrogen burnt. A c.g.s. unit current decomposes in one second 0·000945 gm. of water. The mechanical equivalent of heat being $4·2 \times 10^7$, find in volts the smallest E.M.F. which can decompose water. V.

203. If a mirror galvanometer can just detect 10^{-11} ampere, and if the smallest visible bubble of mixed gases in a water voltameter is a cubic millimetre, how long will it be before a current which just perceptibly deflects the galvanometer will produce perceptible electrolysis ?

204. To what potential would a sphere of radius 10 cms. need to be charged, if by discharging it through a voltameter it is desired to liberate a bubble of hydrogen just perceptible with the aid of a lens, say a millionth of a milligram ? (1 coulomb liberates 0·00001035 gm.)

205. A current of electricity is passed through a wire in a calorimeter, a tangent galvanometer, and a voltameter. When the tangent galvanometer shews a deflection whose tangent is ·7431 the calorimeter gains 5 calories per second and ·00641 gm. of copper is deposited per minute in the voltameter. When the current is increased till ·01282 gm. of copper is deposited per minute, calculate the tangent of the deflection of the galvanometer and the number of calories per second developed in the calorimeter. V.

206. A battery of 5 cells, each of which has an E.M.F. of 2 volts and an internal resistance of ·03 ohm, is connected (a) in series and (b) in parallel, the current passing through a wire of resistance 0·1 ohm. Calculate the heat developed in the wire in each case. V.

207. Two single-needle galvanometers, A and B, are made geometrically similar in all respects, the linear dimensions of A being n times those of B. The magnetic fields are so adjusted by exterior magnets that the periods of vibration of the needles are equal. What is the ratio of the angles of deflection when the same steady current is sent through each?   **L. H.**

208. Find an expression for the magnetic potential of a circular coil of n turns of mean radius r, carrying a current of C amperes, at a point situated on the axis at a distance x from the centre of the coil. Deduce from this an expression for the force exerted by the coil on a magnet pole of m units placed at the point in question. At what point on the axis is the rate of change of this force a maximum for a given small change of x? A coil of 40 turns like that of a tangent galvanometer of mean radius 12 cms. carries a current of 10 amperes. What is the intensity of the field in c.g.s. units which it produces at a point 50 cms. along its axis from the centre of the coil?   **L. H.**

209. A small galvanometer needle, swinging freely under the earth's force alone, makes three oscillations per second. The control magnet of the galvanometer is replaced and adjusted until the needle makes one oscillation in 3 seconds. A millimetre scale is fitted at a distance of a metre from the mirror; find the intensity of the magnetic field at the centre of the galvanometer due to unit current, if a current of 10^{-4} ampere produce a deflection of 20 divisions, the value of H being $0\cdot172$.   **V.**

210. A current traverses a tangent galvanometer of 30 turns of mean radius 8 cms. and produces a deflection of $30°$ at a place where $H = 0\cdot17$. It then traverses a coil of wire of resistance 5 ohms placed in a calorimeter. Calculate the rate of production of heat assuming $J = 42 \times 10^6$. If the deflection increased to 45°, how would the rate of heating be affected?

V.

211. A solenoid of 1000 turns is wrapped uniformly in a single layer on a tube 40 cms. long and 6 cms. in diameter. If the wire carries a current of 0·1 amp., calculate (to three significant figures) the intensity of the field at the middle point of the axis. V. II.

212. Two equal circular coils of wire, each of 10 turns, are mounted parallel to one another and co-axially, and at a distance apart equal to the radius of either. Calculate the intensity of field at the point on the axis midway between the coils, when one ampere flows in the same sense round each coil. V.

213. A square coil of wire of 20 turns and 10 cms. in the side is suspended vertically in the magnetic meridian. What current in amperes must flow through it if the initial torque due to the earth's field be 100 c.g.s. units, the earth's horizontal magnetic field being 0·18 ? V.

214. A vertical ring of radius a carries a current i and is in equilibrium in the earth's field. Find the work necessary to twist it round a vertical axis into the meridian. L.

215. A circular coil of 300 turns of mean radius 10 cms. rotates 10 times per second about a diameter, in the centre of a long solenoid having 5 turns of wire per cm. traversed by a current. The axis of rotation of the coil is perpendicular to the axis of the solenoid. At the instant when the E.M.F. in the coil is a maximum, its ends are connected through a galvanometer to two points A and B on a wire in series with the solenoid, and there is no deflection of the galvanometer. Find in absolute units the resistance between the points A and B. V. II.

216.   A coil rotates about a vertical axis 25 times a second, its radius being 16 cms., and the number of turns in it 313.   A deflection of 39·7 cms. is observed on a scale at a distance of 252 cms. from the deflected magnet which is suspended at the centre of the coil.   Neglecting all corrections, calculate the resistance of the coil.   V. H.

217.   The sides AB, BC, CD, DA of a square have resistances 1, 3, 7, and 5 ohms respectively ; B and D are also joined by a resistance of 2 ohms.   Find the equivalent resistance between A and C.

218.   A framework is made out of six pieces of the same wire forming a square and its two diagonals.   It is fixed together at the angles of the square and at the crossing-point of the diagonals.   What is the equivalent resistance of the frame between the two ends of the same diagonal?   V.

219.   In a Wheatstone net the four arms are 1, 2, 3, 4 ohms respectively, and the galvanometer resistance is 5 ohms. If the battery maintains a P.D. of 1 volt between the corners to which it is joined, find the current in the galvanometer in the two possible cases.

220.   In a Wheatstone net the four resistances are equal and the galvanometer is replaced by a battery of the same E.M.F. as the one already present.   If the resistances in the battery circuits are each the same as those in the other four arms, find the currents in the various branches.   V.

221.   A network is arranged as in Mance's experiment. The resistance of the battery, of the galvanometer, and of each of the other three arms, is 4 ohms, and the E.M.F. of the battery is 2·2 volts ; find the current in the galvanometer (i) with the key open, (ii) when it is closed.   If the resistance of the arm opposite the battery is altered to 5 ohms, find the new values of the current.   L.

222. Two storage cells are connected in parallel to a circuit of 1 ohm resistance. Their E.M.F.'s are 2·05 and 2·15 volts, and their internal resistances ·05 and ·04 ohm respectively. Find the current through each cell.   V. H.

223. A battery of 5 cells in series and a battery of 4 cells also in series have their positive terminals joined together and to one end of a wire of 10 ohms resistance, the other end being joined to the negative terminals of both batteries. If the E.M.F. of each cell is 1 volt, and the resistance of each is 1 ohm, determine the current in each portion of the circuit.

V.

224. A storage cell (2 volts and $\frac{1}{10}$ ohm), a Daniell cell (1 volt and 2 ohms), and a Bichromate cell (1·8 volt and $\frac{1}{2}$ ohm) have their positive terminals all joined, and also their negative terminals. Calculate the current through each cell.

V.

225. Two cells, one a Daniell of E.M.F. 1 volt and resistance 2 ohms, the other a Leclanché of E.M.F. 1·3 volt and resistance 3 ohms, have their positive terminals both joined to one end of a wire of resistance 5 ohms, while their negative terminals are both joined to the other end. Calculate the current in the wire and in each cell.   V.

226. If the vertical magnetic force of the earth be ·4342, calculate the E.M.F. in volts generated in a straight bar 2 metres long carried by a railway train at a rate of 80 kilometres per hour in a direction at right angles to the length of the bar.   V.

227. The axles of the carriages of a train travelling at 60 miles an hour are 5 ft. 3 in. long. Find the difference of potential between their ends, given 1 ft. = 30·5 cms., and the vertical component of the earth's field = 0·41 C.G.S. unit.

V.

228. A river 100 metres broad is flowing northwards along an insulated bed at the rate of 2 metres per sec. ; calculate the difference of potential between the water at the two sides, given that $H = 0\cdot17$ c.g.s. unit and that the magnetic dip is 60°. Does the direction of flow make any difference ?   V.

229. A copper disc of 10 cms. radius spins on its axis (perpendicular to its plane) 3000 times a minute in the earth's horizontal field of force ; find the E.M.F. between the centre and the circumference of the disc. ($H = 0\cdot18$.)   V.

230. A disc of diameter 30 cms. rotates 20 times a sec. in the field due to a very long solenoid of 10 turns per cm. through which 1 ampere is flowing. Find the potential difference between the centre and the circumference of the disc.

231. A circular metal disc of radius 30 cms. rotates on its axis 70 times a second in a uniform magnetic field of intensity 400 c.g.s. units, the disc being normal to the lines of force ; calculate in amperes the current generated when a fixed wire touches the disc at the centre and at the circumference so as to complete a circuit of one ohm electrical resistance.

V.

232. A solid fly-wheel 1 metre in diameter spins on an axis which is directed towards the North at a place where the horizontal intensity is $0\cdot18$. It makes 250 revolutions per min. ; calculate the difference of potential between the centre and the circumference of the wheel.   L.

233. Find the quantity of electricity discharged through a circuit containing a small coil of n turns, when the coil is slipped off from the centre of a long bar magnet of pole-strength μ, the resistance of the whole circuit being R ohms.

V.

234. Find the quantity of electricity induced when a coil of 20 turns, 15 cms. radius, and 4 ohms resistance, lying on a table, is turned over in the earth's magnetic field, the angle of dip being 60° and the value of H being 0·18 c.g.s. unit.

235. Find the total quantity of electricity sent through a galvanometer connected to a coil of ten windings, when the coil is rotated through a right angle about a vertical axis, its final position being in the magnetic meridian. Mean area of coil, 1000 square cms. ; resistance of coil and galvanometer, 1 ohm ; horizontal force of earth's magnetism, 0·17 dyne per unit pole.   V.

236. A flat coil of wire of n turns of mean radius a and mass m has its ends joined and rests on a horizontal table. The density and specific resistance of the material are ρ and σ respectively. Calculate an expression for the quantity of electricity induced when the coil is turned over, given that the horizontal component of the earth's field is H and the dip d.   V.

237. A circular coil 12 cms. in radius and consisting of 150 turns is rotated about a vertical diameter at the rate of 4 revolutions per second. Taking the horizontal component of the earth's magnetic force to be ·18 c.g.s. unit, find the average electromotive force induced in the coil, and also the maximum value of the electromotive force.   V.

238. A wire in the form of a circle 10 cms. in diameter is spinning about a diameter as axis (which is placed vertically) at the rate of 10 revolutions per second in the earth's magnetic field ($H = 0·18$). Calculate the maximum and average E.M.F. generated during a revolution.   V.

239. A straight metal bar of length l is suspended horizontally in the magnetic meridian from a beam by means of two light vertical wires of length L, attached to its extremities, and is set in pendular oscillation in a plane perpendicular to its length, at a region where the earth's field is uniform. Find the magnitude of the E.M.F. generated in the rod and suspending wires at any instant. V. H.

240. A cylindrical tube 50 cms. long and 2 cms. in diameter is uniformly wound with 2000 turns of wire. Calculate the self-induction. V.

241. Calculate roughly the self-induction of 160 metres of wire wound uniformly in the form of a straight solenoid 80 cms. long and 3 cms. in diameter. V.

242. If a wooden curtain ring of mean diameter 21 cms. and sectional diameter 3 cms. be uniformly wound with 1000 turns of very fine wire, calculate its coefficient of self-induction.
L. H.

243. Find the resistance which will be just sufficient to prevent the discharge of a 40 microfarad condenser through a self-induction of half a henry from being oscillatory.
V. H.

244. A condenser is formed by means of two plates having an area of 1 sq. metre at a distance of 1 mm.; it is discharged through a coil having a length of 1 metre and a mean area of 1 sq. cm.; the resistance of the coil is 1000 ohms. Calculate roughly the number of turns in the coil, such that the discharge may just not be oscillatory. V. H.

245. By means of the formula $E = a\,(t_1 - t_2) + b\,(t_1^2 - t_2^2)$, shew that the E.M.F. in a thermoelectric circuit may also be represented as the difference of two terms, each of which involves only one of the temperatures t_1 and t_2 together with the neutral temperature. V.

246. A thermoelectric circuit consists of two equal bars of copper and iron with the ends joined and the junctions maintained respectively at $19.5°$ and $20.5°$ C. Compare the amounts of heat converted into electromagnetic energy and conducted along the bars, given that the thermoelectric power at $20°$ C. $= 18$ microvolts, the specific resistances of copper and iron are 1.6 and 9.8 microhms, and the heat conductivities of the two are 1 and 0.2 c.g.s. unit respectively.　L. H.

247. The thermoelectric power of a copper-lead circuit in microvolts is $-1.34 - 0.0094t$ at temp. t, and for an iron-lead circuit is $-17.2 + 0.048t$. Calculate the E.M.F. in a copper-iron circuit whose junctions are at $50°$ and $250°$ C. respectively.
　　　　　　　　　　　　　　　　　　V. H.

248. The velocity of an ion through air at atmospheric pressure under a potential gradient of 1 volt per cm. is about 1.5 cm. per sec.; assuming that the frictional resistance varies as the velocity, and that the ionic charge is 10^{-19} coulomb, how fast would it move under a force of 1 dyne ?

249. Given that the velocity of a hydrogen ion in electrolysis is 320×10^{-5} cm. per sec. for a potential gradient of 1 volt per cm., find the force necessary to keep 1 gm. of hydrogen moving under similar conditions at a speed of 1 cm. per sec., if an ampere liberates $.00001035$ gm. per sec.

250. Obtain the corresponding result for 1 gm. of potassium (atomic weight $= 39$), given that the velocity of the ion is 66×10^{-5} cm. per sec. for a gradient of 1 volt per cm. (N.B. The forces determined in this and the previous example may be called "ionic friction coefficients," and the student should notice the enormous frictional resistance.)

251. Define the specific molecular conductivity for electricity of a given salt, and shew how it is connected with

the sum of the ionic velocities.   If a solution of **1 gram-**
equivalent in 50 litres has conductivity ·002244, what is the
specific molecular conductivity ?

252.   In an experiment on the electrolysis of silver nitrate,
1·26 gm. of silver is deposited ; also a titration of the solution
round the kathode gave 17·46 gms. of silver chloride before, and
16·68 gms. after, the electrolysis ; if the atomic weights of
silver and chlorine are 108 and 35·5 respectively, determine
the ratio of the velocities of the two ions of silver nitrate.

253.   Find in electrostatic units the charge of electricity
of each kind associated with 1 c.cm. of hydrogen at N.T.P.,
given that a current of 1 ampere passing through acidulated
water decomposes ·00009315 gm. of water per sec., and that a
litre of hydrogen at N.T.P. weighs ·09 gm.

254.   Determine the radius of the drops of water forming
a cloud from the fact that it is observed by means of a micro-
scope to fall at the rate of 0·12 cm. per sec., given that the
velocity of fall is $\frac{2}{9}a^2 g\rho/\mu$ where a is the radius and ρ the
density of the drop, and μ is the viscosity of the air in which
the cloud falls, viz., ·00018 c.ĝ.s. unit.

255.   Ionised air under atmospheric pressure, and saturated
with moisture at 16° C., is suddenly expanded; find the lowest
temperature reached before cloud-formation begins, given that
the height of the barometer is 773 mms., that the saturated
vapour pressure of water at 16° is 13 mms , and that the ratio
of the specific heats of air is 1·41 ; also that when the air is
saturated at its new volume and 16°, the pressure is 572 mms.

256.   Condensation ensues upon the sudden expansion
described in the previous question, and proceeds until the air
is again only just saturated ; shew that the temperature at
which this state of things is reached is 1°·2 C., given that the

density of saturated aqueous vapour at $16°$ is 135×10^{-7} and at $1°·2$ is $51·3 \times 10^{-7}$, while the latent heat of condensation is 606, the specific heat of air at constant volume is ·167, and the mass of 1 c.cm. of air at N.T.P. is ·001293 gm. (The capacity for heat of the aqueous vapour is negligible.)

257. In connection with Qns. 255 and 256, determine the weight of the water-globules in 1 c.cm. of the cloud formed.

258. Assuming that Qns. 254 to 257 relate to the same experiment, and that there are as many water-globules as ions of both kinds, determine the number of ions in 1 c.cm. of air before expansion.

259. By the ionisation of the air in the experiment of Qns. 254 to 258, electricity was caused to leak off a zinc plate $3·2$ cms. in diameter; a quadrant electrometer connected to the plate shewed originally a deflection of 500 divisions, which the leakage reduced by 57 divisions per minute; the space across which the leakage took place had a uniform thickness of $1·2$ cm., and the capacity of the zinc plate and connected quadrants was 62 electrostatic units. From these data and those of Qns. 254 to 258 and from Rutherford's value of the mean ionic velocity in air at N.T.P., viz. $1·5$ cm. per sec. for a potential gradient of 1 volt per cm., calculate the charge on an ion.

260. Shew that if a conducting sphere of radius a carries an electric charge e and is surrounded by a dielectric of specific inductive capacity K, there is a repulsion on the surface of $\dfrac{e^2}{8\pi K a^4}$ per unit area.

261. A soap-bubble of radius a and surface-tension T receives an electric charge e. Assuming that the air inside

A.   4

the bubble obeys Boyle's Law, shew that the new value of the radius x is to be calculated from the equation

$$\left(P + \frac{4T}{a} \right) a^3 = \left(P + \frac{4T}{x} - \frac{e^2}{8\pi K x^4} \right) x^3,$$

where P and K denote the pressure and specific inductive capacity of the gas outside the bubble.

262. Shew how the equation of the preceding example must be modified if at the time of charging the soap-bubble it is contained in a rigid concentric spherical envelope of radius r.

263. Find the numerical solution to Q 261 in the case in which the original radius is 1 cm., the surface tension is 75 dynes per cm., the pressure of the external air is 1000 dynes per sq. cm., and the electric charge is given by connecting the bubble with a source at 30,000 volts.

264. Shew that if the sphere of Q. 57 is electrified with a charge e, the equation becomes

$$P \log_e (P'/P) = \frac{\rho}{\sigma} \left(P' - P + \frac{2T}{a} - \frac{e^2}{8\pi K a^4} \right).$$

265. Find the effect of electric charge on the difference of pressure inside and outside a water-globule of radius ·001 mm., carrying a charge sufficient to raise it to a potential of 4 volts, the surface tension being 75 dynes per cm.

266. What electric charge would neutralise the effect of surface tension in the case of a water-globule of radius (a) 1 cm., (b) ·001 mm., the tension of a surface separating air and water being 75 dynes per cm.? State the result in electrostatic units and also in coulombs in each case. To what potential would the globules be raised respectively?

267. Find the radius of a water-globule of such a size that the quantity of electricity carried by a gaseous ion (about 3.4×10^{-10} electrostatic units, according to J. J. Thomson) would, if placed upon the globule, exactly neutralise the effect of surface-tension upon the vapour pressure outside it, assuming the surface-tension to be 75 dynes per cm. as for large drops. (Note that the result is of the same order as the "radius of molecular action.")

268. Supposing the quantity of aqueous vapour in the air to be $1/\epsilon$ times that which would produce saturation over a flat water-surface, find from the equation of Q. 264 the size of the drops which would be in equilibrium in such an atmosphere, if each drop carried a charge of 3.4×10^{-10} electrostatic units. (Temp. 15° C.)

269. Find to what extent the air in Q. 268 must be dried in order that the drops which are in equilibrium may be half as large as those which would be so over a flat surface.

270. Shew from the equation of Q. 264 that if globules of water carry a charge of 3.4×10^{-10} electrostatic units, then globules of two different sizes will be in equilibrium with any given vapour pressure below a certain limiting value. Find the ratio of this limiting value to that which applies outside a flat surface. Find also the ratio of the corresponding radius to the radius of the globules of Q. 267.

271. Shew from the equation of Q. 264 that if a very small water-globule carry a charge of 3.4×10^{-10} electrostatic units, it cannot grow to a large size unless the air is considerably supersaturated. Determine the amount of supersaturation necessary.

(N.B. *In Qns. 265 to 271, the saturated vapour is assumed to obey Boyle's law, and the surface tension to be the same as for large drops.*)

272. Draw a graph to illustrate Q. 264 shewing how P'/P varies with the radius of the drop, the charge on it being supposed constant. Explain the significance of the form of the graph, especially of the existence of a maximum on it.

273. Shew the bearing of the facts illustrated in the preceding questions on the phenomena of cloud formation; also on our conception of the size of molecules.

274. Assuming that the charge on an atom of hydrogen in electrolysis is the same as that on a gaseous ion, say 3.4×10^{-10} electrostatic units, determine the number of molecules in 1 c.cm. of the gas at N.T.P., given that 1 litre weighs ·09 gm., and that a coulomb liberates ·00001035 gm.

MISCELLANEOUS.

275. A simple pendulum hangs vertically from the roof of a railway carriage travelling at 30 ml./hr. Upon the application of the brake, the pendulum begins to swing to and fro over an arc of 3°. Assuming the resistance of the brake to be constant, find within what distance the train will come to rest.

276. The value of g at a point on the earth's surface is 980 cm./sec.² ; regarding the earth as a uniform sphere of radius 8000 ml., calculate the value of g (i) at a point 10 ml. above the surface, (ii) 10 ml. beneath it.

277. A circular cylinder 1 m. long and of sectional area 300 cm.² is being pushed in the direction of its own axis by a force of 1 megadyne applied at one end. Calculate the internal pressure over the section situated 25 cm. from the front end.   L.

278. A heavy uniform rod of length 100 cm. turns freely on a pivot at a point 30 cm. from one end ; suspended by a string of length 150 cm. fastened to the ends of a rod hangs a bead of the same weight which slides on the string. Graph the potential energy of the system for different inclinations of the rod, and show from the graph that equilibrium is only possible when the rod is vertical.

279. Constructing a similar graph to that required in Q. 278, but taking the distance of the pivot from the end as 40 cm. instead of 30 cm., show that there is equilibrium when the rod is inclined to the vertical at about 61°.

280. Considering the sun and earth to be uniform spheres of radii 430,000 ml. and 4000 ml. respectively, and the mass of the sun to be 324,000 times that of the earth, calculate the gravitational potential energy of the sun. (Mean density of earth = 5·5.)

281. A bullet of mass 1 oz. fired with a velocity of 1600 ft./sec. lodges in the rim of a flywheel (mass, 2000 oz.; radius of gyration, 2 ft.) at a distance of 1·5 ft. from the centre. Find the blow at the axle and the angular velocity set up.

282. A sphere, a thin cylindrical tube, and a solid cylinder roll down a rough inclined plane in the direction of maximum slope. Compare their accelerations. Leeds.

283. Regarding the whole mass (1·5 gm.) of a balance-wheel as concentrated in the rim (diameter 18 mm.), calculate the restoring couple which the watch-spring must exert at a displacement of a radian in order that the period of complete oscillation may be 1 sec. L.

284. A circular disc of wood of uniform thickness and of radius 6 cm. is hung horizontally from its centre by a wire, and the period of torsional oscillation about a vertical axis is 3 sec. Two holes of radius 2 cm. and occupying symmetrical positions on a diameter of the disc are then bored through it and filled with material having 10 times the density of the wood. The period of oscillation at the end of the same wire is now found to be $2\sqrt{5}$ sec.; where are the holes situated?
Leeds.

285. A circular area of radius 4 cm. revolves about a line in its own plane distant 8 cm. from the parallel diameter so as to generate an anchor-ring. Calculate the radius of gyration of the ring about the axis of revolution.

286. Find the effect on the length of a day of transferring 10^8 tons of matter from the North Pole to the Equator, regarding the earth as a uniform sphere of mass 6×10^{21} tons.

287. A top having moment of inertia 500 gm. cm.^{2} about the axis of spin is performing 8 rev./sec. and leaning over at 15° to the vertical. Its mass being 100 gm. and its centre of gravity being at 4 cm. from its point, calculate its rate of precession.

288. A small sphere is hung by a light thread so that its centre is 90 cm. below the point of support; it is drawn aside to a position 20 cm. from the vertical through that point and then released. Find the ratio of its initial period to its period for very small swings.

289. A uniform sphere of radius 2 cm. lies in a spherical bowl of radius 5 cm. and performs small oscillations under the influence of gravity, rolling without slipping. The complete period is observed to be 0·41 sec.; what value does this give for g?

290. A U-tube of uniform section, the limbs of which are long compared with their diameter, is placed with the limbs vertical, and contains enough mercury to fill a length of 50 cm. of the tube; the ends are left open. What is the length of a simple pendulum having the same period as that of the small oscillations of the mercury to and fro in the tube, assuming gravity to be the only appreciable force governing the motion?

291. A hollow sphere of uniform material, whose inner and outer radii are respectively 4 cm. and 8 cm., executes small oscillations as a compound pendulum, swinging about a horizontal axis 6 cm. above the centre of gravity. What is the length of the equivalent simple pendulum?

292. A homogeneous sphere of radius R is suspended by a string of length R; it contains a spherical cavity of radius r whose surface passes through the centre of the sphere, and

the centres of the sphere and cavity are collinear with the string.   Find the length of the equivalent simple pendulum.
Leeds.

293.   A rod of wood of uniform circular section, radius 2 cm., length 20 cm., hangs from an axle situated 5 cm. from the upper end and carries at the lower end a solid metal cylinder of radius 2 cm. and length 5 cm. so disposed as to form a continuation of the rod.   If the metal is 10 times as dense as the wood, find the length of the equivalent simple pendulum.   Leeds.

294.   A compound pendulum has its radius of gyration equal to 35·8 cm. about the line perpendicular to the plane in which its swings are executed.   At what distance from the centre of gravity should the centre of suspension be in order that the period may be (i) least sensitive, (ii) most sensitive, to small displacements of the centre of suspension ?

295.   A uniform heavy rod of length $2l$ oscillates as a compound pendulum about a point in its axis distant x from one end.   Find the time of oscillation under gravity ; also find (if necessary by trial of several values for x/l) about what point the time will be a minimum.   Leeds.

296.   A Kater pendulum of mass 3 kgm. is provided with two knife edges A and B 100 cm. apart ; the c.g. lies between them at a distance of 70 cm. from A, and the moment of inertia about the c.g. is 5×10^6 gm. cm.^{2}.   An adjusting slider of mass 0·5 kgm. can be so attached that its c.g. C shall occupy any desired position along the axis of the pendulum, and its moment of inertia about its own c.g. is 2500 gm. cm.^{2}. What positions of C will make the period of swing about A the same as about B, and why is one of these unsuitable in practice ?   Leeds.

297.   Suppose the mass of the slider in Q. 296 to be varied, its moment of inertia varying in the same proportion ; show

that there is no mass which can under proper conditions bring about the equivalence of period when the position of C is restricted to a range of 40 cm. from the middle point of AB. Find the greatest mass allowable under the same restriction if 80 cm. be substituted for 70 cm. in Q. 296.

298. Replacing the 70 cm. of Q. 296 by 80 cm., and supposing the slider to have mass 200 gm. and moment of inertia 1000 gm. cm.^{2}, find the positions between A and B at which it will make the two periods equal.

299. The slider of Q. 298 is fixed at 40 cm. (i) from A, (ii) from B. Calculate the "frequency of coincidence" of the pendulum so adjusted with a simple pendulum 100 cm. long, the value of g being 981 cm./sec.^{2}.

300. A block of material whose bulk-modulus and shape-modulus are 15×10^{11} and $7 \cdot 5 \times 10^{11}$ dynes/cm.^{2} respectively is subjected to a tensile stress of 81×10^8 dynes/cm.^{2} in one direction and a compressional stress of 162×10^8 dynes/cm.^{2} in another direction at right angles to the first. Calculate the strains in these two directions and in the direction at right angles to both. Leeds.

301. Calculate the strains when the tension of Q. 300 is applied separately; also when a compressional stress of 54×10^8 dynes/cm.^{2} is applied in all three directions; also when this same stress is applied in one direction only, the four parallel faces being held fixed. What stress upon these four faces will be required?

302. Show that the extension due to its own weight of a heavy uniform string hanging vertically is half that produced by an equal weight hung at the end of the string.

303. Find the force necessary to stretch by 1 mm. a rod of iron 1 m. long and 2 mm. in diameter. Assuming that the energy in the stretched rod is converted into heat and imparted to the rod, what rise of temperature will result? (In c.g.s. units Young's modulus = 2×10^{12}, $J = 4\cdot2 \times 10^7$, density of iron = 8, specific heat of iron = 1/9.) Leeds.

304. A metal tube, radius 2 cm., thickness of wall 0·5 mm., has air at a pressure of 100 atmospheres forced into it, and as a result the distance between two marks 100 cm. apart is increased by 0·35 mm. Calculate approximately the bulk-modulus of the material.

305. A brass tube of external and internal diameter 2 cm. and 1·6 cm. respectively and 1 m. long, supported at the ends, is used for determining the traction-modulus by flexure. The modulus is $1\cdot5 \times 10^{12}$ dynes/cm.²; find the sag produced by a load of 1 kgm. placed at the middle. Leeds.

306. A uniform rod rests on two knife-edges at its ends; prove that the sag in the middle is $5l^2/48r$, where l is its length and r the radius of curvature at the middle.

307. Prove that the sag at the middle point of a uniform rod resting on knife-edges at its ends bears to the sag at the ends when the middle only is supported the ratio 5 : 3.

308. A uniform rod of length l is clamped horizontally at each end, the ends being at the same level; where are the points of inflexion?

309. A heavy uniform bar rests across two knife-edges at the same level separated by an interval $2a$, with a length d projecting beyond the knife-edges at either end. For what value of d/a will the bar be horizontal at the points where it is supported? Leeds H.

310. Considering further the problem of Q. 309, state (i) for what minimum value of d the bar will be everywhere concave downwards; also (ii) for what value the middle point will be at the same level as the knife-edges. In the latter case, (iii) where will the points of inflexion be? and (iv) what will the gradient be there and at the knife-edges?

311. A uniform bar of weight W rests across knife-edges situated at 0·215 of its length a from each end. Find the position of (i) the points level with the knife-edges, (ii) the points of inflexion, and (iii) the highest points; also the sag at the middle point and at the ends.

312. A uniform bar is clamped horizontally at one end and rests across a knife-edge placed at the level of the clamp halfway along the projecting portion, which is of length l and weight W. Find the distance from the clamp to (i) the highest point along the bar, (ii) the point of inflexion. Find also the reaction at the knife-edge, the couple exerted by the clamp, the elevation at the highest point of the bar, and the depression at the end.

313. A uniform bar supported as in Q. 312 carries at the outer end a load of 1·2 kgm.; its own weight is 0·6 kgm., its section is a circle of diameter 1 cm., its length is 100 cm., and its traction-modulus is 2×10^{12} dynes/cm.^{2}. Calculate the numerical values of the same quantities as are asked for in Q. 312.

314. A light rod of length $2a$ is clamped horizontally at one end and rests at the other on a knife-edge level with the clamp; find the reaction at the clamp due to a load W hung in the middle; also the reaction at the support. Where is the lowest point of the rod; also, where is the point of inflexion?

315. A capillary tube of uniform bore a metre long is closed at one end and contains air confined by a mercury thread. When the tube is held vertically with the closed end down, the air occupies a length of 50 cm., and the length of the thread is 30 cm., the height of the barometer being 75 cm. If the tube be gently inverted, how much of the thread will remain in the tube?   Leeds.

316. A soap-bubble of radius 5 mm. is contained inside
another whose radius is 2 cm. and the surface tension is
75 dynes/cm. Calculate the pressure inside each bubble,
and also the increase in the radius of the outer bubble if
the inner bursts ; the atmospheric pressure is 10^6 dynes/cm.^{2}.
                                                    Leeds.

317. Two bubbles coalesce ; if V = the consequent change
in volume of the contained air and S = the change in total
surface, show that $3P V = 2ST$, where T is the surface-tension
and P the atmospheric pressure   Leeds.

318. A glass tube of length 20 cm tapers from a diameter
of 3 mm. at one end to 1 mm. at the other, and is held vertically over water (surface tension 80 dynes/cm.) with the
broader end touching the surface. Calculate the rise of the
water in the tube.   L.

319. Two round glass plates of radius 5 cm. are placed
together with a little water between them, which is squeezed
out into a film 0·2 mm. thick occupying practically the whole
of the space between them. What force will be required to
separate them, if the surface tension of water is 80 dynes/cm ?
                                                    Leeds.

320. A piece of wide glass tube is drawn out at one end
until the diameter is 0·5 mm.; it is held vertically and mercury
is gently poured into the wider end. Assuming that a drop
detaches itself when its diameter equals that of the lower end,
calculate to what height the tube may be filled before any
mercury escapes ; the surface-tension of mercury is 540
dynes/cm., and its density 13·6 gm./cm.^{3}.   Leeds.

321. A spherical bubble of oil (radius r) is formed in
a liquid of its own density with which it does not mix ; it
is touched with two equal coaxial circles of wire of radius a
and drawn out into cylindrical form. Prove that the radius

of the spherical ends is $2a$, and that the distance between the wire rings is $4r^3/3a^2 - 2a(16/3 - 3\sqrt{3})$. Also explain why there is a minimum limit to the size of the rings.

Leeds H.

322. A circular shaft 10 cm. in diameter is rotating in a bearing 20 cm. long. Between the shaft and the surface of the bearing there is a film of oil 0·1 mm. thick and of viscosity 0·2. Calculate the couple required to drive the shaft at 1000 rev./min.

Leeds.

323. A vessel of capacity 10 litres contains air at a pressure of three atmospheres; the gas is observed to be leaking at the tap so that the initial rate of escape is 5 cm.³/sec. at atmospheric pressure and temperature. Find how long it will be before the pressure falls to two atmospheres, assuming that the temperature of the whole remains constant. Given $\log_e 2 = 0·7$, $\log_e 3 = 1·1$.

Leeds H.

324. A rod clamped in the middle emits its fundamental note when stroked longitudinally. If the length is 200 cm., the density 7·5, and the frequency of the note produced 1250 per sec., what is Young's modulus for the rod? Leeds.

325. The velocity of sound in water is 1430 m./sec.; calculate the percentage diminution of volume when water is subjected to an additional pressure of one atmosphere, say 10^6 dynes/cm.².

Leeds.

326. If the coefficients of expansion of a crystal along its principal axes are in the ratio of 3, 4 and 5, what number will represent the coefficient along a line equally inclined to the axes?

L.

327. A solid metal sphere weighing 100 gm., of specific gravity 10, is cooled down to $-180°$ C., and immersed in water at $0°$ C. The apparent weight of the sphere in the water with the ice formed on it is then found to be 88 gm.;

calculate the specific heat of the sphere, taking the density of ice to be 0·92 and the latent heat of fusion 80 cal./gm.

Leeds.

328.   Explain the term "corresponding states."   The critical temperature for fluorbenzene is 287° C.; for ether, 194°.   What temperatures for the former correspond to 10° and 33°·8 for ether ?   If at the corresponding temperatures fluorbenzene has a specific volume 1·03 and 1·07 respectively, and if that of ether at 10° is 1·38, what is it according to the theory at 33°·8 ?

329.   The critical temperature for ether is 194° C.; for chloroform, 259°; the critical pressures are 36 and 55 atmospheres respectively.   What temperature for chloroform corresponds to 0° for ether ?   If the compressibility of ether at 0° is ·00011, what is that of chloroform at the corresponding temperature ?

330.   The coefficient of expansion of a perfect gas is 0·003665, its density at N.T.P. is 1·293 mgm./cm.³, and its specific heats at constant pressure and constant volume respectively are 0·2374 and 0·1690 cal./gm.   Calculate the heat absorbed per unit change of volume while a gram of this gas expands isothermally under approximately normal pressure.

331.   The vapour pressure of water at 10°, 12°, and 14° is 9·17, 10·46, 11·91 mm. of mercury respectively.   Calculate the mass of water in a litre of saturated air at these three temperatures, Boyle's Law being supposed applicable to the saturated vapour.

332.   The density of saturated aqueous vapour at 10°, 12° and 14° is 9·356, 10·600 and 11·987 mgm. per litre respectively ; show that if a litre of saturated air at 10° be mixed with a litre at 14° in a 2-litre vessel impervious to heat a certain amount of condensation will result.   Explain the bearing of this upon the subject of cloud-formation.

333. Determine the velocity of mean square for air-molecules at N.T.P., given that the densities of air and mercury are 0·001293 and 13·6 respectively.   Leeds.

334. Water at the rate of 0·8 gm. per sec. flows through a glass tube 50 cm. long with a platinum wire along its axis. If the wire is heated by a current of 0·2 amp. under a pressure of 16 volts, find how much the water is warmed in its passage through the tube, the value of J being $4·2 \times 10^7$ ergs/cal.

L.

335. Pouillet found that the earth receives from the sun in the form of radiant energy 1·76 cal. per cm.^{2} per min. Assuming the sun's energy of radiation to arise from the diminution of its gravitational potential energy resulting from contraction, calculate the period for which a contraction of 1 ml. in the sun's radius would provide the necessary energy. (The sun's distance is 9×10^7 ml.; for other data see Q. 280.)

336. Calculate the absolute temperature of melting ice from the following data: the Joule-Thomson effect for a certain gas is 0°·037 C. per atmosphere of 10^6 dynes/cm.^{2}, its relative expansion from 0° to 100° is 0·36613, its specific heat at constant pressure is 3·41 cal./gm., and its specific volume at 0° and 760 mm. is 11·1 litres/gm.

Leeds H.

337. The U-tube of Q. 290 is now closed at the two ends, which isolates a length of 70 cm. of air at atmospheric pressure in one limb and 50 cm. in the other. If the barometric height is 75 cm., and the ratio of the specific heats of air is 1·4, find the length of the equivalent simple pendulum under the new conditions, the changes being supposed adiabatic.

338. In an experiment of Müller's, carried out by Assmann's method as indicated in Q. 290 and Q. 337, the periods of oscillation were 0·631 sec. and 0·543 sec. respectively.

The barometer read 76·1 cm., the volumes of air confined were 1180 cm.^{3} and 1160 cm.^{3} respectively, and the mean section of the tube was 3·97 cm.^{2}. What value do these figures give for the ratio of the specific heats?   Leeds H.

339. Assuming that steam obeys the gas laws and that the specific volume is 1640 cm.^{3}/gm. at 100°, find roughly the heat-equivalent of the work done on 1 gm. of saturated steam at 100° C. (vapour pressure 76 cm.) in bringing it to saturation at 101° (v.p. 78·7 cm.); a pressure of 76 cm. of mercury may be taken as 10^6 dynes/cm.^{2}, and $J = 42 \times 10^6$ ergs/cal. From the result and the fact that the specific heat of steam at constant pressure is about 0·48, deduce that the specific heat of saturated steam is negative.

340. Find approximately the specific heat of saturated steam at 100° C. and at 200°, if the rate of increase of the total heat of saturated steam be taken as constant and equal to 0·3 per 1° C.

341. Dry atmospheric air at the surface of the earth (temp. 30° C.) is enclosed in a vessel impervious to heat, conveyed to a height of 100 m., and then suddenly released. What must be the atmospheric temperature at that height in order that the released air shall tend neither to rise nor to sink? (The height of the homogeneous atmosphere at the time of the experiment is 8850 m.; the ratio of the specific heats of dry air is 1·4.)

342. Make an approximate calculation similar to that in Q. 341 for air saturated with moisture at 30° C. with the help of the following data : saturation pressure at 30°, 31·51 mm.; at 29°, 29·74 mm.; barometric height at surface of earth, 76 cm.; density of mercury, 13·6 gm./cm.^{3}; $g = 981$ cm./sec.^{2}; $J = 4·2 \times 10^7$ ergs/cal.; 1 gm. of dry air at N.T.P. occupies 771 cm.^{3}; 1 gm. of saturated vapour occupies 34·90 litres

at 29°, 33·05 at 30°; latent heat of vaporisation at 30°, 579 cal./gm.; specific heat of air at constant volume, 0·169 cal./gm.

343.   A 20 H.P. non-condensing engine is supplied with steam at 150° C. and requires 120 lb. of coal per hr.; a 10 H.P. engine is supplied with steam at 140° C. and requires 61 lb. of coal per hr.   Taking into account the maximum attainable efficiency of each engine, determine which is the more nearly perfect.   L.

344.   A locomotive with a non-condensing engine of ideal efficiency uses 1200 lb. of coal per hr. when overcoming frictional resistance of 1·5 tons at a steady rate of 40 ml./hr. on level ground.   The calorific value of 1 lb. of coal is 8650 pound-calories, and 1 pound-calorie = 1400 ft.-lb.   Find the efficiency of the engine and the temperature of the steam in the boiler.   Leeds.

345.   A vessel contains water (refractive index 4/3) to a depth of 3 cm., and a plate of glass 2 cm. thick (refractive index 3/2) is held so that one face just touches the water-surface.   What will be the apparent depth of the bottom of the vessel below the upper face of the plate?   Leeds.

346.   A lens is placed with one face against a flat mirror and it is found that the point of a pin at 10 cm. from the lens coincides with its own image.   The experiment is repeated with the space between the lens and mirror occupied by water (refractive index 4/3), and the new distance is 16 cm.   Find the radius of curvature of the wetted surface.   Leeds.

347.   What must be the distance between two thin convergent coaxial lenses of focal lengths 21 cm. and 20 cm. that parallel rays falling on the first may have their final focus 12 cm. from the second?   Will the image be erect or inverted?
                                                                         Leeds.

A.   5

348. A telescope has an objective of 15 cm. focal length; its eyepiece is of the Huyghens type, the focal lengths of the components being 7·5 cm. and 2·5 cm., and the separation being 5 cm. Show that when focused for infinity there will be 11·25 cm. between the objective and the field-lens, that the magnifying power will be 4, and the eyepiece equivalent to a single lens of focal length 3·75 cm. Leeds.

349. In an optical system the first medium is air and the last water (refractive index 4/3); the distance from the first focal point F to the first nodal point N is 12 cm.; how far is the second focal point f from its nodal point n, and where are the unit planes? Leeds.

350. In the system of Q. 349 where are the points from which distances must be measured in order that the usual formula $1/v + 1/u = 1/F$ may hold good?

351. A system resembling that of Q. 349 reversed has $F'N'$ equal to 6 cm. and $n'f'$ equal to 8 cm.; it is placed in line with the other so that the points occur in the order $FNnf$, $F''N'n'f''$; also $fF' = 16$ cm. Where are the focal points ϕ_1, ϕ_2 and the nodal points ν_1, ν_2 of the combination? What is the nature of each system and of the combination?

352. Newton's rings are formed between a lens and a flat surface and are viewed normally by transmitted light. If the radius of curvature of the face of the lens be 50 cm. and if the squares of the radii of the first six bright rings are 23, 52, 81, 111, 141 and 174 times 10^{-4} cm.^{2}, make the best calculation you can of the wave-length of the light. Leeds.

353. Compare the resolving power of a diffraction grating 2 in. broad ruled with 3000 lines per inch with that of a $1\frac{1}{2}$ in. grating ruled with 10,000 lines per inch; also compare the separation of the two sodium lines in the first spectrum, the same magnification being applied to each.

354. The focal lengths of the objective and eyepiece of a telescope are 6 ft. and 0·5 in. respectively. The aperture is stopped down to a rectangle measuring 4 in. by 2 in. What will be the shape and size of the "image" of a star in minutes of arc subtended at the eye, the wave-length of the light being considered to be 550 $\mu\mu$? Leeds H.

355. A Rochon double image prism is composed of two 45° prisms of quartz, of which the refractive indices are 1·553 and 1·544; when parallel light falls normally on the prism, what will be approximately the angle between the two beams in the prism and after emergence?

356. Two right-angled isosceles prisms of spar (refractive indices 1·658 and 1·486) with their edges parallel to the optic axis are fastened together with an air-film between to form a Foucault prism; find the angular aperture of the polarised beam in the plane of symmetry of the compound prism.

357. The specific rotation of quartz for the B line ($\lambda = 687\mu\mu$) is 15° 30', and for the G line ($\lambda = 431\mu\mu$) 42° 20'; a piece of quartz 4 cm. thick with faces normal to the optic axis is viewed between (i) crossed Nicols, (ii) parallel Nicols, by white light with the aid of a spectroscope. How many black bands will lie between B and G in the spectrum, and what will be approximately the corresponding wave-lengths?

358. The three principal refractive indices of nitre being 1·3346, 1·5056, 1·5064, what is the angle between the optic axes?

359. In a biaxal crystal the angle between the optic axes is 60°, while that between the directions of single ray-velocity is 90°; in what proportion are the principal wave-velocities in the crystal?

360. Determine the proportion of the principal wave-velocities in a crystal when the diameter of the circle of contact of the tangent plane at the extremity of the optic axis with the wave-surface equals the axis, the inclination of the ray-axis to the optic axis being 30°.

361. Calculate the capacity of a sphere of diameter 15 cm. inside which there is an earthed concentric sphere of diameter 10 cm. L.

362. Find the capacity of a parallel plate condenser, area A, separation of plates d, when a slab of dielectric of thickness t and specific inductive capacity K is placed between the plates. L.

363. An air condenser is formed of two concentric spherical conductors of radii 10 and 11 cm. Calculate the energy spent in charging the condenser by a battery of 500 storage cells, each of which has an E.M.F. of two volts.

Leeds.

364. A sphere of radius 1 cm. contains a dilute solution of sodium chloride. Supposing so many chlorine ions removed as correspond to 10^{-5} gm. of salt, calculate the electric intensity (in volts per cm.) at the surface of the sphere, taking the equivalent for hydrogen to be 0·0000104 gm./coulomb, and the atomic weights of sodium and chlorine to be 23 and 35·5 respectively. L.

365. For steady potentials of 50, 60, 70 volts respectively an electrostatic voltmeter gives readings of 15, 18, and 21 divisions ; what will be the indication corresponding to an alternating E.M.F. $70 \sin pt$, assuming the frequency $p/2\pi$ to be rapid in relation to the movement of the voltmeter?

L.

366. ABC is a wire triangle, the resistance of the sides AB, BC, CA being 1, 2 and 3 ohms respectively ; each of the corners is joined to a fourth point D by a wire of resistance 4 ohms. Calculate the resistance between A and B.

Leeds.

367. $ABCDA$ is a network of conductors whose resistances are as follow: $AB = 2$, $BC = 3$, $CD = 4$, $DA = 5$, $BD = 6$ ohms. Calculate the resistance between A and C.   Leeds.

368. $ABCD$ is a wire quadrilateral with a connection across AC; the side AB and the diagonal AC contain cells of E.M.F. 1·1 and 1·2 volts respectively, the positive terminals being toward A; if the resistances of the arms AB, BC, CD, DA, AC are 3, 2, 1, 2, 3 ohms respectively, calculate the current in each cell.   Leeds.

369. Two electrolytic resistances of 5 ohms and 10 ohms are joined in parallel, and a current is sent through them from a battery of E.M.F. 8 volts and internal resistance 2 ohms. The E.M.F. of polarisation in the respective tubes is 0·1 volt and 1·8 volts; find the currents transmitted.   L.

370. A battery of E.M.F. 6 volts and internal resistance 0·5 ohm is joined in parallel with another of E.M.F. 10 volts and resistance 1 ohm, and the combination is used to send current through an external resistance of 12 ohms. Find the current in each cell.   L.

371. Convert an inductance of 1 henry into electrostatic mgm.-mm.-min. units.   Leeds.

372. Express 1 megohm, 1 microfarad, and 1 henry in electrostatic c.g.s. units.   Leeds.

373. The capacity of a condenser is found by the leakage method to be 2 sec. per megohm; how does it compare with a spherical air-condenser whose radii are 10·0 and 10·1 cm.?   Leeds.

374. The self-inductance of a coil of resistance 100 ohms is 0·5 henry; find its equivalent resistance for alternating currents of frequency (i) 100, (ii) 200 per sec.   Leeds.

ANSWERS.

1. 746 watts.

4. $x = \dfrac{gt}{k} + \dfrac{(\epsilon^{-kt} - 1)}{k}\left(\dfrac{g}{k} - V\right)$, where x and t are the distance and time reckoned from the position where the velocity was V. Terminal velocity, $\dfrac{g}{k}$.

8. H.P. $= 38190$; time $= 127\cdot6$ hrs.; resistance $= 236\cdot24$ tons' weight; coal $= 4353$ tons. (*Geographical* miles are meant.)

9. $6\cdot81 \times 10^{-12}$ kgm.-weight.   10. 3762 secs.

11. Mass of sun : mass of earth $= (390)^3 : (13)^2$.

12. $11\cdot2$ lbs. per ton; $358\cdot4$ H.P.

13. 4400 ft.-lb.-sec. units per sec.; 17600 ft.-poundals per sec.

14. $140\cdot15 : g$.   15. $\dfrac{2\pi^2}{g}$.   16. $\frac{3}{4}$.

17. $1\frac{0}{2}\frac{0}{7}\frac{0}{6}^3 g$ cms. per sec.

18. $\sqrt{480/g}$ secs.; $\sqrt{15g/2}$ radians per sec.

19. $96\frac{3}{17}$ cms. per sec. per sec.

20. If the mass M be suspended at an arm r, and the moment of inertia of the wheel and axle be I, the acceleration $= \dfrac{Mr^2 g}{Mr^2 + I}$. If a length l unwrap in time t, the friction couple is
$$Mrg - \left(Mr + \dfrac{I}{r}\right)\dfrac{2l}{t^2}.$$

21. $\sqrt{\dfrac{3g}{a}}$.

22. When the rod has turned through an angle θ, the two amounts are $\frac{1}{4} mgl \sin\theta$ and $\frac{3}{8} mgl \sin\theta$ respectively.

23. Let the added mass have a radius of gyration k, and let its centre of gravity be x below the knife-edge; then the positions are given by $\dfrac{k^2+x^2}{x} <=> \dfrac{g}{\pi^2}$ respectively.   25. (i) $a \cdot \dfrac{\sqrt{3}}{2}$; (ii) $a \cdot \dfrac{5\sqrt{3}}{12}$.

26. (i) $a \cdot \dfrac{2\sqrt{2}}{3}$ below the point of suspension; (ii) $\dfrac{5a}{6}$ below.

27. Becomes $\sqrt{1\cdot2}$ times as long.   28. $52\cdot88$ cms.

29. $2\pi \sqrt{\dfrac{D}{g}}$.

30. (i) Multiplied by $4\sqrt{2}$; (ii) divided by $2\sqrt{2}$; (iii) doubled.

31. $\frac{27}{100}$ sec.   32. $2\cdot8 \times 10^7$ pounds' weight per sq. in.

33. $2\cdot5 \times 10^9 g$ dynes per sq. cm.   34. Halved.

35. $\dfrac{9\pi g}{25600}$ cm.   36. $0\cdot1875$ cm.   37. $\frac{10000}{3} g$ ergs per c. cm.

38. Torsional rigidity, about $2\cdot23 \times 10^{10} g$ dynes per sq. cm.; simple rigidity, $2/\pi$ times as much.

39. Five per cent., if the errors in n and Y are in opposite senses.

40. $22001 : 22010\cdot7$.   41. $0\cdot000474$ gm.

42. $\dfrac{10000}{7g}$ cms., assuming that the liquid wets the tube.

43. $2\cdot88 \times 10^8$; $\dfrac{2880}{g}$ cms.   44. $1\cdot2$ cm.

45. 296000 dynes per sq. cm.   46. $0\cdot0809$ cm.

47. $0\cdot1764$ poundal per ft.   48. $\dfrac{3g}{40}$ dynes per cm.   49. $\dfrac{2TA}{x}$.

50. $(76s + 10) g + 156000$, where s is the density of mercury.

51. 125960 ergs.   52. $\frac{1}{16}$ gm. weight per sq. cm.   53. $\dfrac{36}{17g}$ cm.

54. $\dfrac{P'+24}{0\cdot2214P}$, where P' is the external pressure and P is a normal atmosphere, both in dynes per sq. cm.

58. (a) $0\cdot0192$, $1\cdot016$; (b) $19\cdot2$, 1016 dynes per sq. cm.

60. (a) $1\cdot152 \times 10^{-7}$ cm., better $1\cdot662 \times 10^{-7}$; (b) $1\cdot016 \times 10^{-6}$, better $1\cdot466 \times 10^{-6}$.

61. $0\cdot00077$ cm.   62. 226 cms.

63. Let the pressure outside the tyre be one atmosphere, and let the internal pressure be n atmospheres; then the time taken is about $863000 \ (n^2-1) \log_{10} \left(\dfrac{n^2+n-2}{n^2-n-2} \right)$ seconds. If $n=2$, this is infinite (as it should be); if $n=3$, the time is $2 \cdot 75 \times 10^6$ secs.; for $n=4$, $3 \cdot 30 \times 10^6$ secs.; for $n=10$, $7 \cdot 60 \times 10^6$; for $n=20$, $15 \cdot 05 \times 0^6$. When the internal pressure considerably exceeds the external (as was probably intended), a rougher calculation is appropriate, and leads to the result $750000 \ n$ secs. It will be noticed that at 20 atmospheres, this is practically correct.

64. $1:17$.   65. Observer at rest, $19:21$; observer on train, $361:441$.

66. $1086\frac{2}{3}\frac{3}{3}$; 1080.   67. $1 \cdot 34 \times 10^{12}$ dynes per sq. cm.

68. $4 \cdot 032 \times 10^{11}$ dynes per sq. cm.

69. $2^{19} \times 17000$ poundals per sq. ft.

70. $10^9 g$ dynes per sq. cm.; $\sqrt{\dfrac{10^9 g}{3}}$ cms. per sec.

71. Frequency $\dfrac{1}{2} \ \sqrt{\dfrac{1100 g}{7}}$.

72. $\dfrac{10}{3} \ \sqrt{\dfrac{50 g}{77\pi}}$.   74. 120, taking $g=980$ cms. per sec. per sec.

75. $\frac{101000}{8}$ cms. per sec.   76. 33047 cms. per sec.

77. $70\frac{5}{6}$ cms.   78. $\frac{125}{7157}$ c. cm.   79. $-\frac{1}{10000}$.

80. The temperature-coefficient $=-2 \cdot 47 \times 10^{-4}$.

81. $1 \cdot 0011$; $\cdot 9989$.   82. $75 \cdot 604$ cms.

83. 1 metre $=1 \cdot 09376$ yd.   84. $4 \cdot 75 \times 10^{-5}$.   85. $30 \cdot 02278$ gms.

86. $\frac{23}{180}$ of the internal volume should be filled.

87. Force exerted per degree $=$ coefficient of expansion multiplied by force necessary to double the length of the bar (upon the usual understanding).

88. Gains $5 \cdot 646$ secs. per day for each degree rise of temperature.

89. About $10 \cdot 2$ secs.   90. About $288° \mathrm{C}$.   91. $0 \cdot 0536$.

92. $1 \cdot 402$.   93. $0 \cdot 2384$.   94. 6000.

95. $0 \cdot 00001259$.   96. $1 : 0 \cdot 365 : 4 \cdot 34$.   97. $8 : 1$; equal.

98. (i) $3:1$; (ii) $243:35$.   99. $21° \cdot 15$.   100. $1 \cdot 194$ gm.

101. $13° \mathrm{C}$.   102. $30 \cdot 55$, $49 \cdot 71$, $81 \cdot 86$ cms. respectively.

103. 0·3424.   104. $100 \sqrt{15sg}$, where s is the density of mercury.

105. 46612 cms. per sec.   106. $8·5 \times 10^{-6}$ cm.

107. If θ is the original absolute temperature, then the new temperature is the antilog. of $\{\log \theta + (\gamma - 1) \log 10\}$, where γ is the ratio of the specific heats.

108. (i) 467° C.; (ii) 288°·5 C., taking $\gamma = 1·41$.

109. (a) 50 atmospheres, (b) 239 atmospheres; (a) no rise, (b) 1032° C.

110. $-125°·6$ C., taking $\gamma = 1·41$.   111. 2184° C.

112. $\frac{7}{83}$ calories per gm.   113. $2·02 \times 10^9$ ergs per gm.

114. JL/g, where J is Joule's equivalent, L the latent heat of fusion, and g the acceleration of gravity.

115. About 26300 cms. per sec.

116. 0°·885 C., but the question is faulty, the data being inconsistent.

117. Internal, $20·8 \times 10^9$ ergs; external, $1·71 \times 10^9$.

118. 0·0634.   119. About 2·99 times as much as is needed.

120. Gain of 0·00368 units.

121. Loss of 1·75 units accompanied by a (larger) gain which the data do not determine.   122. $\dfrac{6·95 \times 10^7}{JL}$ lbs., J being in ft.-lbs. per calorie.

123. (b) is $12\frac{3}{4}$ times as good as (a).   124. 1672 c. cms.

125. Falls 0°·3755 C.   126. 0°·0252 C.   127. $-\frac{1}{2888}$.

128. Let c be the capacity for heat (in ergs) of the substance of the bubble, θ its initial (absolute) temperature, and A the increase in area; then if surface tension varies uniformly with temperature, the difference in work is $\theta (0·17A - c)$ very approximately.

129. Mean interval $\pm 15·29$ secs.   130. 48 hrs. $\pm 17·21$ secs.

131. 2 kilometres per sec.

132. Away from the earth with $\frac{2}{1833}$ of the velocity of light.

133. 25 cms. below the true centre.

134. For a Huyghens eye-piece, focal lengths 2 ins. and $\frac{2}{3}$ in.; for a Ramsden eye-piece, two lenses each of focal length $\frac{4}{3}$ in.

135. (a) 2 or $\frac{1}{2}$; (b) 2 or $\frac{1}{2}$; (c) 200 or $\frac{1}{200}$.

136. (i) 1 : 100 : 576, (ii) 1 : 1 : 0·0576. The higher power.

137. The first. $21\frac{9}{13}$ cms. convex, $21\frac{9}{17}$ cms. concave.

138. 10·85 ins. convex, 17·00 ins. concave.   139. $p_1 + p_2 + p_1 p_2 d$.

140. 20 cms.; infinity. **141.** $18\frac{3}{4}$ ins. **142.** 1·5.

143. 6 ins. **144.** $2\frac{7}{17}$ ins.

145. $\frac{140}{40}$ cm. and $\frac{100}{3}$ cm. from faces.

146. (i) $3\frac{1}{4}$ ins.; $2\frac{7}{17}$, $3\frac{1}{4}$. (ii) $2\frac{10}{23}$ ins.; $2\frac{9}{25}$, $2\frac{9}{25}$. (iii) $5\frac{5}{8}$ ins.; $6\frac{3}{8}$, $4\frac{3}{8}$.

147. 3·6 ins.; 3 ins. and 4 ins.; 1·5.

148. (i) Water against flat surface; nodal points at $1\frac{1}{17}$ and $\frac{7}{18}$ inch from curved and flat surfaces; focal points at $3\frac{1}{4}$ and $3\frac{1}{2}$ ins. from the same. (ii) Water against curved surface; the corresponding values are $3\frac{1}{4}$, $3\frac{11}{12}$, 13, and $9\frac{1}{17}$.

149. (i) $8\frac{6}{8}$ or $8\frac{7}{8}$ ins. according as the object is on the side of the air or the water. (ii) The corresponding values are $27\frac{1}{3\cdot7}$ and $8\frac{1}{4\cdot17}$ ins. Magnification : (i) $\frac{19}{11}$ or $\frac{88}{23}$; (ii) $\frac{1}{7}$ or $\frac{1}{3}\frac{1}{7}$.

150. 0·0398 sq. cms.

151. Nodal points at $\frac{6}{8}$ and $\frac{3}{8}$ inch from the faces in contact with air and water respectively; focal points at $3\frac{1}{8}$ and $4\frac{1}{2}$ ins.

152. (i) $13\frac{3}{13}$ ins. (ii) $16\frac{1}{4}$ ins. Magnification: (i) $\frac{23}{13}$; (ii) $\frac{7}{4}$.

153. $\dfrac{1·41}{100}$ inch; 7·5 ins. **154.** 4 metres.

155. $\frac{1}{15000}$ of a radian. **156.** 0·0135 cm. **157.** 0·0295 cm.

158. $\dfrac{0·001525}{\lambda}$ cm. **159.** 85·2 cms. **160.** 5×10^{-5} cm.

161. $3·535 \times 10^{-5}$ cm. **162.** $2·654 \times 10^{-5}$ cm.

163. 1·01836 mm. **164.** $\frac{22}{000}$. **165.** 108.

166. At all points on two surfaces of revolution having an asymptotic cone of semi-vertical angle $54\frac{3}{4}°$ with its apex at the middle point of the magnet.

167. (i) The two points on the horizontal section of the above surfaces where the force due to the magnet is equal and opposite to H. (ii) On the meridian, at a distance r from the centre given by the equation $2Mr = H(l^2 - r^2)^2$; where M and $2l$ denote the moment and length of the magnet.

168. 1157 c.g.s. units. **169.** 0·98. **170.** 0·0468 per cent.

171. If the magnetic moments are M and M', and the distance between the centres is d, the couples are $\dfrac{MM'}{d^3}$ and $\dfrac{2MM'}{d^3}$, both in the same sense.

172. 109 : 416.

173. If the length of the line joining the centres is d, the field is either $8\sqrt{2} \cdot \dfrac{M}{d^3}$ at right angles to that line, or $16\sqrt{2} \cdot \dfrac{M}{d^3}$ along it, or zero.

174. If d is the distance between the centres, the field is $8\sqrt{5}\,\dfrac{M}{d^3}$ in a direction inclined to the line of centres at $\cot^{-1} 2$.

175. If the magnets are at distances a and b from the point, the field is $\dfrac{2M\sqrt{a^6+b^6}}{a^3 b^3}$ inclined at an angle $\tan^{-1}\left(\dfrac{a^3}{b^3}\right)$ to the distance a.

176. $\dfrac{M}{a^3}(13\cdot5 \pm 6\sqrt{3})$ along the direction of the third magnet.

177. (i) $\dfrac{3\sqrt{3}}{2} \cdot \dfrac{M}{a^3}$ along the median, or $\dfrac{1}{2}\dfrac{M}{a^3}$ at right angles to it;

(ii) $\dfrac{27}{2}\dfrac{M}{a^3}$ along, or $\dfrac{15\sqrt{3}}{2}\dfrac{M}{a^3}$ at right angles.

178. (i) $2\sqrt{3} \cdot \dfrac{M}{a^3}$ along the median or $\dfrac{2M}{a^3}$ at right angles to it;

(ii) $\dfrac{9}{2}\dfrac{M}{a^3}$ along, or $\dfrac{21\sqrt{3}}{2}\dfrac{M}{a^3}$ at right angles.

179. $\dfrac{2}{5}\sqrt{\dfrac{g}{5}}$ electrostatic units. **180.** $\dfrac{A}{4\pi d}$; $\dfrac{1}{d}$.

181. $\sqrt{\dfrac{g}{20\pi}}$ c.g.s. units. **182.** $44\cdot2$ ergs.

183. 451 grams' weight.

184. 100890 volts, taking an atmosphere as 10^6 dynes per sq. cm.

185. $10\cdot8$ cms.; $3\frac{1}{3}$ units.

186. If $Q =$ the charge on the particle and $Q' =$ that on the sphere of radius a, then the limiting distance d is given by $\dfrac{d^2(d+2a)^2}{a(d+a)^3} = \dfrac{Q}{Q'}$.

187. $8\cdot70 \times 10^7$ ohms. **188.** $(a)\ \dfrac{25}{3\times10^9}$, $(b)\ \dfrac{25}{3\times10^8}$, $(c)\ \dfrac{25}{3\times10^7}$.

189. $\dfrac{1}{36\pi}$. **190.** 10. **191.** $0\cdot829$.

192. Electrical : coal $= 3730 : 99$.

193. 36×10^{12} ergs; 10 amperes; 2 volts.

194. $4\cdot91 \times 10^{-11}$ penny; 20357 kilometres.

195. $0\cdot294$ watt; $1\cdot764$ volt. **196.** 4160 watts; 2048 watts.

197. 71·22 ohms.　　　　　　　**198.** 0·318 units.

199. 10J absolute units.　　　**200.** 0·913 $\frac{p}{V}$ pence.

201. 0·3402 gm.; 142·3 calories, 69·7 calories, 52·3 calories.

202. 1·508 volt.　　　　**203.** $18\frac{3}{8}$ years.

204. $8·70 \times 10^6$ volts (quite impracticable, of course).

205. 1·4862 ; 20 calories per sec.

206. (a) 38·1,　(b) 8·48 calories per sec.

207. $\delta_A : \delta_B = 1 : n^3$, the two needles being magnetised with equal intensity.

208. $2\pi nC \left\{ 1 - \dfrac{x}{\sqrt{r^2 + x^2}} \right\}$; $\dfrac{2\pi nCr^2m}{(r^2 + x^2)^{\frac{3}{2}}}$; $x = \dfrac{r}{2}$; 0·2662 units.

209. $\frac{17}{81}$ c.g.s. units.　　**210.** 0·002066 calories per sec.; 0·006198.

211. 3·11 c.g.s. units.

212. $\dfrac{8·992}{r}$, where $r =$ the radius of either coil.

213. $\frac{3}{9}$ amperes.　　**214.** $\pi a^2 iH$.　　**215.** $3·72 \times 10^8$ cms. per sec.

216. $3·085 \times 10^{10}$ cms. per sec.　　**217.** $\frac{6}{23}$ ohms.

218. A piece of the same wire of length $(2 - \sqrt{2})$ times the side of the square.

219. $\frac{1}{81}$ ampere or $\frac{1}{85}$ ampere.

220. Let R be the resistance of each arm, and E be the E.M.F. of each battery; then the two batteries and two opposite arms each transmit a current $\dfrac{1}{2} \dfrac{E}{R}$, while the other two arms carry no current.

221. (a) $1\frac{1}{5}$ amp. in each case. (b) Open, $\frac{9.9}{1.0.0}$ amp.; closed, $\frac{9.9}{7.4.0}$ amp

222. $\frac{1.1.5}{8.4.4}$ and $\frac{9}{1.0}$ amp. respectively.

223. $\frac{5}{11}$ amp. in the larger battery, $\frac{3}{11}$ in the smaller, $\frac{4}{11}$ in the wire.

224. 0·72, 0·464, 0·256 amp. respectively.

225. $\frac{2.6}{1.5.5}$ amp., $\frac{3}{8.2}$ amp., $\frac{4.1.5}{3.1.5}$ amp. respectively.

226. 0·00193 volt.　　　　**227.** 0·00176 volt.

228. 0·00589 volt.　　**229.** $\dfrac{9\pi}{10^5}$ volt.　　**230.** $\dfrac{18\pi^2}{10^5}$ volt.

231. 0·792 ampere.　　　　**232.** $5·89 \times 10^{-5}$ volt.

233. $\dfrac{4\pi\mu n}{R} \times 10^{-8}$ coulomb.   234. $2 \cdot 204 \times 10^{-5}$ coulomb.

235. $1 \cdot 7 \times 10^{-5}$ coulomb.   236. $\dfrac{mH \tan d}{2\pi n\sigma\rho}$.

237. $0 \cdot 001954$ volt; $0 \cdot 00307$ volt.   238. $\dfrac{9\pi^2}{10^7}$ volt; $\dfrac{18\pi}{10^7}$ volt.

239. When the wires are inclined at θ to the vertical, the E. M. F. is
$2lV \cos\theta \sqrt{gL\left(\sin^2\dfrac{a}{2} - \sin^2\dfrac{\theta}{2}\right)}$, where V is the vertical component of the earth's field, and a is the maximum angular displacement.

240. $0 \cdot 00316$ henry.   241. 32×10^{-4} henry.

242. $0 \cdot 001346$ henry.   243. $223 \cdot 6$ ohms.   244. 4194 turns.

246. $9 : 1596000$.   247. $0 \cdot 00145$ volt.

248. $1 \cdot 5 \times 10^{12}$ cms. per sec.   249. $3 \cdot 02 \times 10^{14}$ dynes.

250. $3 \cdot 83 \times 10^7$ kilograms' weight.

251. Sum of velocities $= EV_x \times$ sp. mol. cond., where E is the electrochemical equivalent of hydrogen, and V_x is the potential gradient.
$112 \cdot 2 \dfrac{\text{sq. cms.}}{\text{ohm gm.}}$.

252. $1 : 1 \cdot 146$.   253. $1 \cdot 3 \times 10^{10}$ units.   254. $3 \cdot 15 \times 10^{-4}$ cm.

255. $-18° \cdot 2$ C.   257. $4 \cdot 80 \times 10^{-6}$ gm.   258. $5 \cdot 0 \times 10^4$.

259. $7 \cdot 8 \times 10^{-10}$ electrostatic units.   263. $1 \cdot 111$ cm.

265. Decreases it by $\dfrac{1}{675\pi}$ of itself.

266. (a) $61 \cdot 4$ units, $2 \cdot 05 \times 10^{-8}$ coulomb; 18420 volts. (b) $6 \cdot 14 \times 10^{-5}$ units, $2 \cdot 05 \times 10^{-14}$ coulomb; $184 \cdot 2$ volts.

267. $3 \cdot 13 \times 10^{-8}$ cm.   268. $2 \cdot 90 \times 10^{-8}$ cm.

269. The vapour pressure must be reduced to $8 \cdot 6 \times 10^{-23}$ times the saturation-pressure over the flat surface.

270. $5 \cdot 55 : 1$; $1 \cdot 587 : 1$.

271. At least $5 \cdot 55$ times the normal saturation-pressure.

274. About $3 \cdot 84 \times 10^{19}$.

275. About 384 yd.   276. $977 \cdot 55$, $978 \cdot 78$ cm./sec.²

277. $833\frac{1}{3}$ dynes/cm.²

278. Taking as datum-line the level of the pivot, and varying the inclination by $10°$ at a time from the vertical, the successive values of the negative potential energy are proportional to $35·0$, $35·1$, $35·4$, $36·1$, $37·4$, $38·8$, $41·2$, $44·8$, $49·7$, $55·9$, $63·5$, $72·2$, $81·2$, $90·2$, $98·6$, $105·3$, $110·6$, $113·9$, 115.

279. The new values are $55·0$, $54·8$, $54·2$, $53·4$, $52·7$, $51·7$, $51·2$, $51·7$, $53·1$, $55·9$, $60·1$, $67·1$, $71·2$, $77·3$, $83·3$, $88·0$, $91·8$, $94·2$, $95·0$.

280. $2\frac{1}{4} \times 10^{48}$ ergs.　　　　**281.** 100 lb. ft./sec. forward; $0·3$ radian/sec.

282. $30 : 21 : 28$.　　　　　　**283.** 48 dyne cm.

284. Their centres are 3 cm. from the centre of the disc.

285. $8·72$ cm.　　　　**286.** Longer in the ratio $(1 + \frac{5}{12} . 10^{-13}) : 1$.

287. About 4 radians/sec.　　　　　**288.** $1·00315 : 1$.

289. 986 cm./sec.^{2}.　　　　**290.** 25 cm.　　　　**291.** $10·72$ cm.

292. $(22 R^5 - 20 R^2 r^3 \pm 20 R r^4 - 7 r^5)/(10 R^4 - 10 R r^3 \pm 5 r^4)$.

293. $17·1$ cm.　　　　**294.** (i) $35·8$ cm.; (ii) 91 cm.

295. $2\pi \sqrt{\{(4 l^2 - 6 l x + 3 x^2)/3 g (l - x)\}}$; minimum when $x/l = 0·423$.

296. 70 cm. from A in direction BA, or $71·38$ cm. on either side of the middle of AB; the first brings the c.o. to the middle of AB.

297. $223·5$ gm.　　　　**298.** $38·67$ cm. from the middle point.

299. With A uppermost, (i) one every 177 sec., (ii) 179 sec.; with B uppermost, (i) 50 sec., (ii) 47 sec.

300. 66×10^{-4}, extension; 96×10^{-4}, contraction; 12×10^{-4}, extension.

301. 42×10^{-4}, extension, and 12×10^{-4}, contraction; 12×10^{-4}, contraction; $21·6 \times 10^{-4}$, contraction; $21·6 \times 10^8$ dynes/cm.^{2}.

303. 64 kgm.; $0°·0268$ C.　　　　**304.** $1·9 \times 10^{12}$ dynes/cm.^{2}.

305. $0·294$ mm.　　**308.** $0·289 l$ from the middle.　　**309.** $\sqrt{2/3}$.

310. (i) $d = a$; (ii) $d = a \sqrt{5/6}$; (iii) at $a/\sqrt{6}$ on either side of the middle; (iv) $\pm w a^3/(18 I E \sqrt{6})$, $\mp w a^3/(12 I E)$, where w is the weight of unit length, I the "moment of inertia" of cross-section, and E the traction-modulus.

311. (i) $0·454 a$; (ii) $0·324 a$; (iii) $0·353 a$, all measured from the middle point; $0·00044 W a^3/I E$, $0·00815 W a^3/I E$, below the knife-edges.

312. (i) $0.349l$; (ii) $0.195l$; (iii) $17W/16$; (iv) $Wl/32$; (v) $0.00084Wl^3/IE$; (vi) $0.0143Wl^3/IE$.

313. (i) 33.5 cm.; (ii) 16.9 cm.; (iii) weight of 3.638 kgm.; (iv) 31.9 kgm. cm.; (v) 0.61 mm.; (vi) 9.60 mm.

314. Force $11W/16$ and couple $3Wa/8$; $5W/16$; $1.106a$ and $0.545a$ from the clamp.

315. Nearly 14 cm.

316. Greater than atmospheric pressure by 150 and 750 dynes/cm.^{2} respectively; about 6×10^{-6} cm.

318. 1.13 cm.   **319.** 642 gm. wt.   **320.** 3.24 cm.

322. $10\pi^2/3$ megadyne cm.   **323.** 53 m. 20 s.

324. 1.87×10^{12} dynes/cm.^{2}.   **325.** 0.0049.   **326.** $4\sqrt{3}$.

327. 0.102 cal./gm.   **328.** $66°.4$ C. and $94°.9$; 1.43.

329. $38°$ C.; 0.000072.   **330.** 0.0241 cal./cm.^{3}.

331. 9.43, 10.68, 12.07 mgm. respectively.   **333.** 485 m./sec.

334. $0°.952$.   **335.** Over 50 years.   **336.** $273°.00$.

337. 8.93 cm.   **338.** 1.36.   **339.** 1.3 cal.

340. -1.14; -0.68 cal./gm.   **341.** $29°.02$ C.

342. Between $29°.63$ and $29°.64$, according to the mode of approximation.

343. The ratio for comparison is $61 : 75$.   **344.** 4.9%; $118°$ C.

345. $3\frac{7}{17}$ cm.   **346.** $8\frac{2}{3}$ cm.   **347.** 28.5 cm.; inverted.

349. 9 cm.; at 9 cm. from F and 12 cm. from f respectively.

350. $6\sqrt{3}$ cm. from each focal point.

351. $f'\phi_2 = 3$ cm., $\phi_2\nu_2 = 4.5$ cm., $\phi_1 F = 6.75$ cm., $\nu_1\phi_1 = 4.5$ cm.; the separate systems are convergent; the combination is divergent, of equivalent focal length 4.5 cm.

352. 6×10^{-5} cm.   **353.** $2 : 5$; $3 : 10$.

354. A rectangular patch 5.4 by 10.7.

355. About $20'$ and $31'$.   **356.** About $9°.2$.

357. (i) 639, 577, 531, 494, 464, 439; (ii) 679, 606, 552, 511, 479, 451 $\mu\mu$.

358. About $7°$.   **359.** $\sqrt{6} : \sqrt{3} : \sqrt{2}$.

360. $b/a = 0.659, c/a = 0.318$.   **361.** 22.5 cm.

362. $KA/\{4\pi (Kd - Kt + t)\}$.   **363.** $611\frac{1}{2}$ ergs.   **364.** 1.48×10^{10}.

365. 14.8 divisions.   **366.** $\frac{1}{1}\frac{4}{3}\frac{1}{1}$ ohm.   **367.** $\frac{1}{1}\frac{3}{3}\frac{1}{1}$ ohm.

368. $\frac{1}{13}$ amp., $\frac{3}{130}$ amp.   **369.** 1.03 amp., 0.345 amp.

370. $\frac{3}{8}\frac{4}{7}$ amp., $\frac{1}{3}\frac{0}{7}\frac{0}{7}$ amp.   **371.** 3.09×10^{-17}.

372. $1.11 \times 10^{-6}, \; 9 \times 10^{5}, \; 1.11 \times 10^{-12}$.   **373.** $180,000 : 101$.

374. 380 ohms, 686 ohms.

PRINTED IN ENGLAND BY J. B. PEACE, M.A.
AT THE CAMBRIDGE UNIVERSITY PRESS